Modulation
Transfer Function
in Optical and
Electro-Optical
Systems

Tutorial Texts Series

Modulation Transfer Function in Optical and Electro-Optical Systems

GLENN D. BOREMAN

Tutorial Texts in Optical Engineering
Volume TT52

SPIE PRESS
A Publication of SPIE—The International Society for Optical Engineering
Bellingham, Washington USA

Library of Congress Cataloging-in-Publication Data

Boreman, Glenn D.
    Modulation transfer function in optical and electro-optical systems / Glenn D. Boreman.
        p. cm. — (Tutorial texts in optical engineering ; v. TT 52)
    Includes bibliographical references and index.
    ISBN 0-8194-4143-0
    1. Optics. 2. Electrooptical devices. 3. Modulation theory. I. Title. II. Series.

TA1520 .B67 2001
621.36–dc21   2001031158
                                                                                    CIP

Published by

SPIE—The International Society for Optical Engineering
P.O. Box 10
Bellingham, Washington 98227-0010 USA
Phone: 360/676-3290
Fax: 360/647-1445
Email: spie@spie.org
WWW: www.spie.org

Printed in the United States of America.

TABLE OF CONTENTS

PREFACE

I first became aware that there was such a thing as MTF as an undergraduate at Rochester, scurrying around the Bausch and Lomb building. There was, in one of the stairwells, a large poster of the Air Force bar target set. I saw that poster every day, and I remember thinking…gee, that's pretty neat. Well, more than 25 years later, I still think so. I have had great fun making MTF measurements on focal-plane arrays, SPRITE detectors, scanning cameras, IR scene projectors, telescopes, collimators, and infrared antennas. This book is an outgrowth of a short course that I have presented for SPIE since 1987. In it, I emphasize some practical things I have learned about making MTF measurements.

I am grateful for initial discussions on this subject at Arizona with Jack Gaskill and Stace Dereniak. Since then, I have had the good fortune here at Central Florida to work with a number of colleagues and graduate students on MTF issues. I fondly recall discussions of MTF with Arnold Daniels, Jim Harvey, Didi Dogariu, Karen MacDougall, Marty Sensiper, Ken Barnard, Al Ducharme, Ofer Hadar, Ric Schildwachter, Barry Anderson, Al Plogstedt, Christophe Fumeaux, Per Fredin, and Frank Effenberger. I want to thank Dan Jones of the UCF English Department for his support, as well as Rick Hermann, Eric Pepper, and Marshall Weathersby of SPIE for their assistance and enthusiasm for this project. I also appreciate the permissions granted for reproductions of some of the figures from their original sources.

Last but surely not least, I want to thank Maggie Boreman – my wife, main encourager, and technical editor. Once again, Meg, you have wrestled with my occasionally tedious exposition and transformed it, if not into poetry, then at least into prose. Thanks.

GDB
Cocoa Beach
15 March 2001

CHAPTER 1
MTF IN OPTICAL SYSTEMS

Linear-systems theory provides a powerful set of tools with which we can analyze optical and electro-optical systems. The spatial impulse response of the system is Fourier transformed to yield the spatial-frequency optical transfer function. Simply expressing the notion of image quality in the frequency domain does not by itself generate any new information. However, the conceptual change in viewpoint − instead of a spot size, we now consider a frequency response − facilitates additional insight into the behavior of an imaging system, particularly in the common situation where several subsystems are combined. We can multiply the individual transfer function of each subsystem to give the overall transfer function. This procedure is easier than the repeated convolutions that would be required for a spatial-domain analysis, and allows immediate visualization of the performance limitations of the aggregate system in terms of the performance of each of the subsystems. We can see where the limitations of performance arise and which crucial components must be improved to yield better overall image quality.

In Chapter 1, we develop this concept and apply it to classical optical systems, that is, imaging systems alone without detectors or electronics. We will first define terms and then discuss image-quality issues.

1.1 Impulse response

The impulse response $h(x,y)$ is the smallest image detail that an optical system can form. It is the blur spot in the image plane when a point source is the object of an imaging system. The finite width of the impulse response is a result of the combination of diffraction and aberration effects. We interpret $h(x,y)$ as an irradiance (W/cm^2) distribution as a function of position. Modeling the imaging process as a convolution operation (denoted by *), we express the image irradiance distribution $g(x,y)$ as the ideal image $f(x,y)$ convolved with the impulse response $h(x,y)$:

$$g(x,y) = f(x,y) * h(x,y) \ . \tag{1.1}$$

The ideal image $f(x,y)$ is the irradiance distribution that would exist in the image plane (taking into account the system magnification) if the system had perfect image quality, in other words, a delta-function impulse response. The ideal image is thus a magnified version of the input-object irradiance, with all detail preserved. For conceptual discussions, we typically assume that the imaging system has unit magnification, so that we can directly take $f(x,y)$ as the object irradiance distribution, albeit as a function of image-plane coordinates. We can see from Eq. (1.1) that if $h(x,y) = \delta(x,y)$, the image is a perfect replica of the object. It is within this context that $h(x,y)$ is also known as the point-spread function (PSF). A perfect optical system is capable of forming a point image of a point object. However, because of the blurring effects of diffraction and aberrations, a real imaging system has an impulse response that is not a point. For any real system $h(x,y)$ has finite spatial extent. The narrower the PSF, the less blurring occurs in the image-forming process. A more compact impulse response indicates better image quality.

As Fig. 1.1 illustrates, we represent mathematically a point object as a delta function at location (x',y') in object-plane coordinates

$$f(x_{\text{obj}},y_{\text{obj}}) = \delta(x'-x_{\text{obj}},y'-y_{\text{obj}}) \;. \tag{1.2}$$

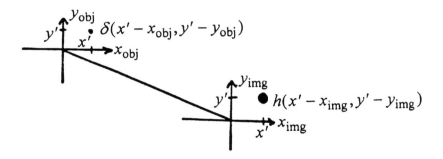

Figure 1.1 A delta function in the object is mapped to a blur function, the impulse response, in the image plane.

Assuming that the system has unit magnification, the ideal image is a delta function located at (x', y') in image-plane coordinates

$$g(x_{\text{obj}},y_{\text{obj}}) = \delta(x'-x_{\text{img}},y'-y_{\text{img}}) \;. \tag{1.3}$$

In a real imaging system, instead of a delta function at the ideal image point, the impulse response is centered at $x' = x_{\text{img}}$ and $y' = y_{\text{img}}$ in the image plane $g(x_{\text{img}},y_{\text{img}}) = h(x'-x_{\text{img}},y'-y_{\text{img}})$, in response to the delta-function object of Eq. (1.2). We represent a continuous function $f(x_{\text{obj}},y_{\text{obj}})$ of object coordinates, by breaking the continuous object into a set of point sources at specific locations, each with a strength proportional to the object brightness at that particular location. Any given point source has a weighting factor $f(x', y')$, which we find using the sifting property of the delta function:

$$f(x', y') = \iint \delta (x' - x_{obj}, y' - y_{obj}) f(x_{obj}, y_{obj}) \, dx_{obj} \, dy_{obj} \; . \qquad (1.4)$$

The image of each discrete point source will be the impulse response of Eq. (1.1) at the conjugate image-plane location, weighted by corresponding object brightness. The image irradiance function $g(x_{img}, y_{img})$ becomes the summation of weighted impulse responses. This summation can be written as a convolution of the ideal image function $f(x_{img}, y_{img})$ with the impulse response

$$g(x_{img}, y_{img}) = \iint h(x' - x_{img}, y' - y_{img}) f(x_{img}, y_{img}) \, dx' dy' \; , \qquad (1.5)$$

which is equivalent to Eq. (1.1). Figure 1.2 illustrates the imaging process using two methods: the clockwise loop demonstrates the weighted superposition of the impulse responses and the counterclockwise loop demonstrates a convolution with the impulse response. Both methods are equivalent.

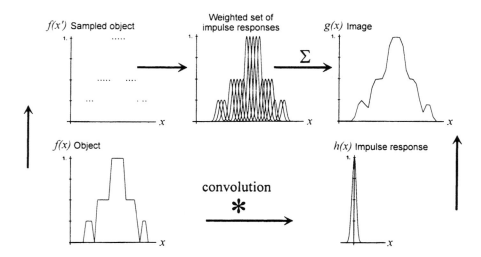

Figure 1.2 Image formation can be modeled as a convolutional process. The clockwise loop is a weighted superposition of impulse responses and the counterclockwise loop is a convolution with the impulse response.

Representing image formation as a convolutional process assumes linearity and shift invariance (LSI). To model imaging as a convolutional process, we must have a unique impulse response that is valid for any position or brightness of the point-source object. Linearity is necessary for us to be able to superimpose the individual impulse responses in the image plane into the final image. Linearity requirements are typically accurately satisfied for the irradiance distribution itself (the so-called aerial image). However, certain detectors such as photographic film, detector arrays (especially in the IR), and xerographic media are particularly nonlinear in their impulse response. In these cases, the impulse response is a function of the input irradiance level. We can only perform LSI analysis for a restricted range of input irradiances. Another linearity consideration is that coherent optical systems (optical processors) are linear in electric field (V/cm), while incoherent systems (imaging systems) are linear in irradiance (W/cm^2). We will deal exclusively with incoherent imaging systems. Note that partially coherent systems are not linear in either electric field or irradiance and their analysis—as a convolutional system—is more complicated, requiring definition of the mutual coherence function.[1]

Shift invariance is the other requirement for a convolutional analysis. According to the laws of shift invariance, a single impulse response can be defined that is not a function of image-plane position. Shift invariance assumes that the functional form of $h(x,y)$ does not change over the image plane. This shift invariance allows us to write the impulse response as $h(x'-x_{img}, y'-y_{img})$, a function of distance from the ideal image point, rather than as a function of image-plane position in general. Aberrations violate the assumption of shift invariance because typically the impulse response is a function of field angle. To preserve a convolutional analysis in this case, we segment the image plane into isoplanatic regions over which the functional form of the impulse response does not change appreciably.

1.2 Spatial frequency

We can also consider the imaging process from a frequency-domain (modulation-transfer-function) viewpoint, as an alternative to the spatial-domain (impulse-response) viewpoint. An object- or image-plane irradiance distribution is composed of "spatial frequencies" in the same way that a time-domain electrical signal is composed of various frequencies: by means of a Fourier analysis. As seen in Fig. 1.3, a given profile across an irradiance distribution (object or image) is composed of constituent spatial frequencies. By taking a one-dimensional profile across a two-dimensional irradiance distribution, we obtain an irradiance-vs-position waveform, which can be Fourier decomposed in exactly the same manner as if the waveform was in the more familiar form of volts vs time. A Fourier decomposition answers the question of what frequencies are contained in the waveform in terms of spatial frequencies with units of cycles (cy) per unit distance, analogous to temporal frequencies in cy/s for a time-domain waveform. Typically for optical systems, the spatial frequency is in cy/mm.

An example of one basis function for the one-dimensional waveform of Fig. 1.3 is shown in Fig. 1.4. The spatial period X (crest-to-crest repetition distance) of the waveform can be inverted to find the x-domain spatial frequency denoted by $\xi \equiv 1/X$.

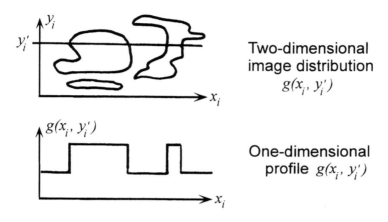

Two-dimensional image distribution $g(x_i, y_i')$

One-dimensional profile $g(x_i, y_i')$

Figure 1.3 Definition of a spatial-domain irradiance waveform.

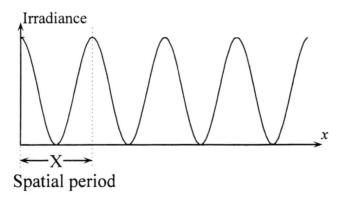

Irradiance

x

←X→
Spatial period

Figure 1.4 One-dimensional spatial frequency.

Fourier analysis of optical systems is more general than that of time-domain systems because objects and images are inherently two-dimensional, and thus the basis set of component sinusoids is also two-dimensional. Figure 1.5 illustrates a two-dimensional sinusoid of irradiance. The sinusoid has a spatial period along both the x and y directions, X and Y respectively. If we invert these spatial periods we find the two spatial frequency components that describe this waveform: $\xi = 1/X$ and $\eta = 1/Y$. Two pieces of information are required for specification of the two-dimensional spatial frequency. An alternate representation is possible using polar coordinates, the minimum crest-to-crest distance, and the orientation of the minimum crest-to-crest distance with respect to the x and y axes.

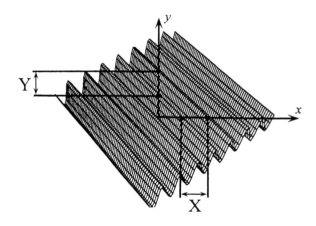

Figure 1.5 Two-dimensional spatial frequency.

Angular spatial frequency is typically encountered in the specification of imaging systems designed to observe a target at a long distance. If the target is far enough away to be in focus for all distances of interest, then it is convenient to specify system performance in angular units, that is, without having to specify a particular range distance. Angular spatial frequency is most often specified in cy/mrad. It can initially be a troublesome concept because both cycles and milliradians are dimensionless quantities but, with reference to Fig. 1.6, we find that the angular spatial frequency ξ_{ang} is simply the range R multiplied by the target spatial frequency ξ. For a periodic target of spatial period X, we define an angular period $\theta \equiv X/R$, an angle over which the object waveform repeats itself. The angular period is in radians if X and R are in the same units. Inverting this angular period gives angular spatial frequency $\xi_{ang} = R/X$. Given the resolution of optical systems, often X is in meters and R is in kilometers, for which the ratio R/X is then in cy/mrad.

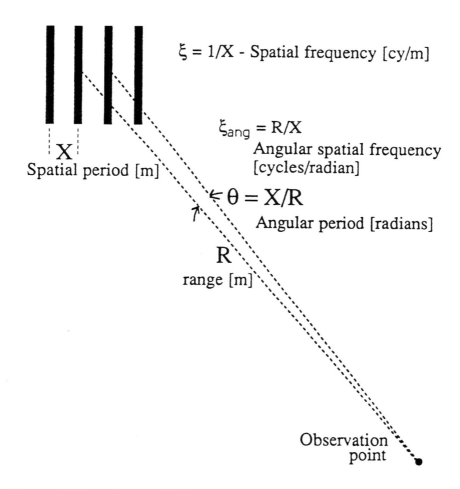

$\xi = 1/X$ - Spatial frequency [cy/m]

$\xi_{ang} = R/X$
Angular spatial frequency
[cycles/radian]

$\theta = X/R$
Angular period [radians]

Figure 1.6 Angular spatial frequency.

1.3 Transfer function

Equation (1.1) describes the loss of detail inherent in the imaging process as the convolution of the ideal image function with the impulse response. The convolution theorem[2] states that a convolution in the spatial domain is a multiplication in the frequency domain. Taking the Fourier transform (denoted $\mathcal{F}$) of both sides of Eq. (1.1) yields

$$\mathcal{F}[g(x,y)] = \mathcal{F}[f(x,y) * h(x,y)] \tag{1.6}$$

and

$$G(\xi,\eta) = F(\xi,\eta) \times H(\xi,\eta) \ , \tag{1.7}$$

where uppercase functions denote the Fourier transforms of the corresponding lowercase functions: F denotes the object spectrum, G denotes the image spectrum, and H denotes the spectrum of the impulse response. $H(\xi,\eta)$ is the transfer function, in that it relates the object and image spectra multiplicatively. The Fourier transform changes the irradiance waveform from a spatial-position function to the spatial-frequency domain, but generates no new information. The appeal of the frequency-domain viewpoint is that the multiplication of Eq. (1.7) is easier to perform and visualize than the convolution of Eq. (1.1). This convenience is most apparent in the analysis of imaging systems consisting of several subsystems, each with its own impulse response. As Eq. (1.8) demonstrates, each subsystem has its own transfer function as the Fourier transform of its impulse response.

The final result of all the subsystems operating on the input object distribution is a multiplication of their respective transfer functions. Figure 1.7 illustrates that we can analyze a combination of several subsystems by the multiplication of transfer functions of Eq. (1.9) rather than the convolution of impulse responses of Eq. (1.8):

$$f(x, y) * h_1(x, y) * h_2(x, y) * \ldots * h_n(x, y) = g(x, y) \qquad (1.8)$$

and

$$F(\xi,\eta) \times H_1(\xi,\eta) \times H_2(\xi,\eta) \times \ldots \times H_n(\xi,\eta) = G(\xi,\eta) \;. \qquad (1.9)$$

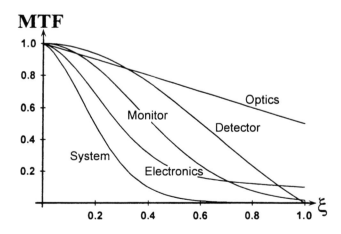

Figure 1.7   The aggregate transfer function of several subsystems is a multiplication of their transfer functions.

For the classical optical systems under discussion in this first chapter, we ignore the effects of noise and we typically assume that $H(\xi,\eta)$ has been normalized to have unit value at zero spatial frequency (a uniform image-irradiance distribution). This normalization yields a relative transmittance for the various frequencies and ignores attenuation factors that are independent of spatial frequency, such as Fresnel reflections or material absorption. Although this normalization is common, when we use it, we lose information about the absolute signal levels. For some cases we may want to keep the signal-level information, particularly when electronics noise is a significant factor.

With this normalization, $H(\xi,\eta)$ is referred to as the optical transfer function (OTF). Unless the impulse response function $h(x,y)$ satisfies certain symmetry conditions, its Fourier transform $H(\xi,\eta)$ is in general a complex function, having both a magnitude and a phase portion, referred to as the modulation transfer function (MTF) and the phase transfer function (PTF) respectively:

$$\text{OTF} \equiv H(\xi,\eta) = \left| H(\xi,\eta) \right| \exp[-j\theta(\xi,\eta)] \qquad (1.10)$$

and

$$\text{MTF} \equiv \left| H(\xi,\eta) \right| \qquad \text{PTF} \equiv \theta(\xi,\eta) \ . \qquad (1.11)$$

1.3.1 Modulation transfer function

The modulation transfer function is the magnitude response of the optical system to sinusoids of different spatial frequencies. When we analyze an optical system in the frequency domain, we consider the imaging of sinewave inputs (Fig. 1.8) rather than point objects.

Figure 1.8 Sinewave target of various spatial frequencies.

A linear shift-invariant optical system images a sinusoid as another sinusoid. The limited spatial resolution of the optical system results in a decrease in the modulation depth M of the image relative to what is was in the object distribution (Fig. 1.9). Modulation depth is defined as the amplitude of the irradiance variation divided by the bias level:

$$M = \frac{A_{\max} - A_{\min}}{A_{\max} + A_{\min}} = \frac{2 \times ac \text{ component}}{2 \times dc \text{ component}} = \frac{ac}{dc} \quad . \qquad (1.12)$$

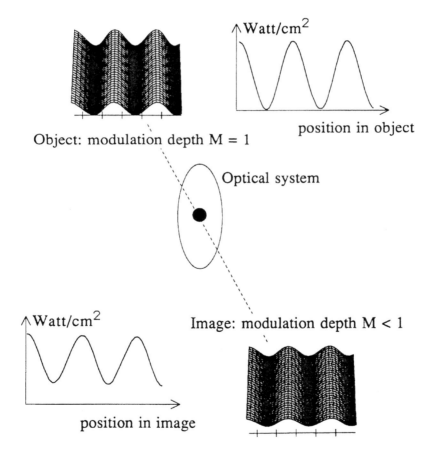

Figure 1.9 **Modulation depth decreases going from object to image.**

We can see from Fig. 1.10 that modulation depth is a measure of contrast, with

$$M \to 0 \text{ as } A_{\max} - A_{\min} \to 0 \text{ and } M \to 1 \text{ as } A_{\min} \to 0. \qquad (1.13)$$

The $M = 0$ condition means that, although there is still a nonzero image-irradiance level, there is no spatial variation of that irradiance. When the waveform has a minimum value of zero, there is unit modulation depth—whatever the maximum irradiance level is. Low levels of modulation depth are harder to discern against the unavoidable levels of noise inherent in any practical system.

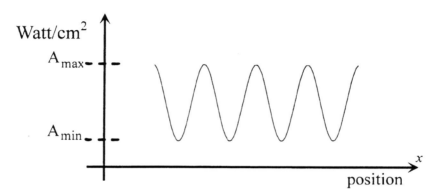

Figure 1.10 Definition of modulation depth as described in Eq. (1.10).

The finite size of the impulse response of the optical system causes a filling in of the valleys of the sinusoid and a lowering of the peak levels. The effect of this is to decrease the modulation depth in the image relative to that in the object. Defining the modulation transfer (MT) as the ratio of modulation in the image to that in the object

$$MT \equiv M_{\text{image}}/M_{\text{object}} \qquad (1.14)$$

we find that the reduction of modulation transfer is spatial-frequency dependent. The limited resolution of the optics is more important at high spatial frequencies, where the scale of the detail is smaller. As seen in Fig. 1.11, when we plot modulation transfer against spatial frequency, we obtain the MTF, generally a decreasing function of spatial frequency. The MTF is the image modulation as a function of spatial frequency (assuming a constant object modulation),

$$MTF(\xi) \equiv M_{\text{image}}(\xi)/M_{\text{object}} . \qquad (1.15)$$

The M of the object waveform does not need to be unity – if a lower input modulation is used, then the image modulation will be proportionally lower.

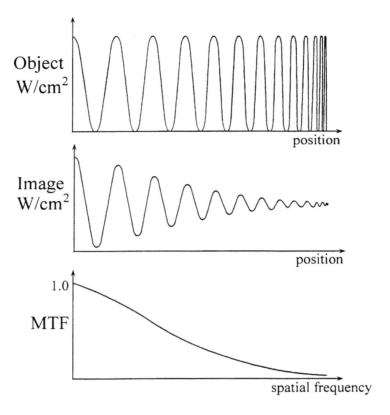

Figure 1.11 Modulation transfer function is the decrease of modulation depth with increasing frequency.

1.3.2 Phase transfer function

Recalling the definition of the optical transfer function in Eq. (1.10), we now proceed to the interpretation of the phase response:

$$\text{OTF}(\xi) \equiv \text{MTF}(\xi) \exp\{-j\,\text{PTF}(\xi)\} \ . \qquad (1.16)$$

For the special case of a symmetric impulse response centered on the ideal image point, the phase transfer function (PTF) is particularly simple, having a value of either zero or π as a function of spatial frequency (see Fig. 1.12). The OTF can be plotted as a bipolar curve having both positive and negative values. If the MTF were plotted on the same graph, it would be strictly above the ξ axis. The PTF shows a phase reversal of π over exactly that frequency range for which the OTF is below the axis.

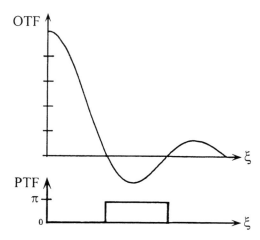

Figure 1.12 OTF and PTF corresponding to a defocused impulse response.

We see the impact of this phase reversal on the image of a radial spoke target that exhibits increasing spatial frequency toward the center (Fig. 1.13). The image was blurred at a 45° diagonal direction. Over a certain range of spatial frequencies, what begins as a black bar at the periphery of the target becomes a white bar, indicating a phase shift of π. This phase shift occurs in a gray transition zone where the MTF goes through a zero.

Other phase transfer functions are possible. For instance, a shift in the position of the impulse response from the ideal geometric image position produces a linear phase shift in the image. Typically this displacement is a function of field angle. If the impulse point-spread function (PSF) is asymmetric such as seen for a system with the aberration coma, the PTF will be a nonlinear function of frequency (Fig. 1.14). We do not usually measure the phase transfer function directly; rather we calculate it from a Fourier transform of the impulse response. Most often, PTF is used as an optical-design criterion, to determine the presence and severity of certain aberrations.

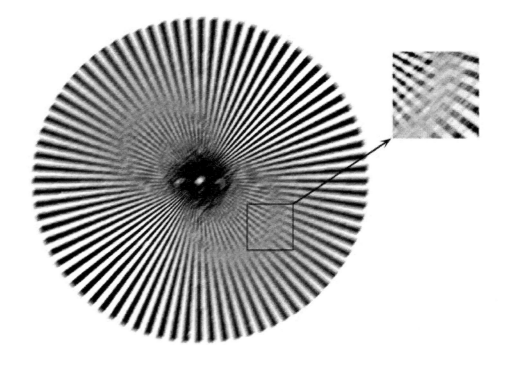

Figure 1.13 Image of spoke target that has been blurred in the 45° diagonal direction. Phase reversals exist over certain spatial-frequency ranges, and are seen as white-to-black line transitions.

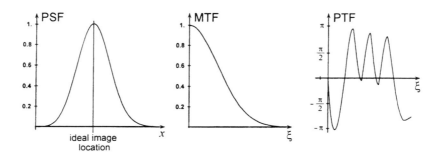

Figure 1.14 Asymmetric impulse responses produce nonlinear PTFs.

Nonlinearities in the PTF, also called phase distortion, cause different spatial frequencies in the image to recombine with different relative phases. The result of phase distortion is a change in the shape of the spatial waveform describing the image. Severe phase distortion produces image waveforms that are dramatically different from the object waveform. To illustrate the effects of this PTF nonlinearity, we consider the imaging of four-bar targets with equal lines and spaces. Although these targets are not periodic in the true sense (being spatially limited), we will speak of nth harmonics of the fundamental spatial frequency as simply $n\xi_f$. In Fig. 1.15, we plot PTF and image-irradiance waveforms for three π-phase step-discontinuity PTFs. To emphasize the effect of the PTF on the image, we let MTF = 1 in this example, so that all of the spatial frequencies are present in their original amounts and only the relative phases have changed.

In case I (Fig. 1.15), the PTF transition occurs at $4\xi_f$. There is no shift for the fundamental or the third harmonic, and a π shift for higher frequencies. We see a phase-reversal artifact as a local minimum at the center of each bar, primarily because the fifth harmonic is out of phase with the third and the fundamental. In case II, the transition occurs at $2\xi_f$, so the only in-phase spatial frequency is the fundamental. The bars are sinusoidal at the center, with secondary-maxima artifacts in the shoulder of each bar and in the space between them, arising primarily from the third and fifth harmonics. The step transition for case III occurs at $0.9\xi_f$, shifting the fundamental and all harmonics with respect to frequencies lower than ξ_f. The most dramatic artifact is that the image now has five peaks instead of the four seen in cases I and II.

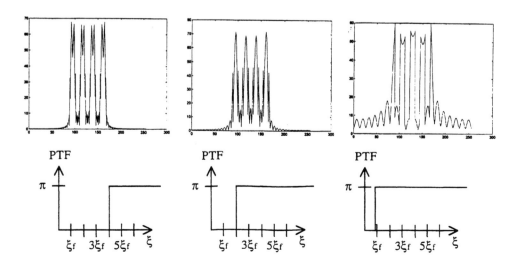

Figure 1.15 Image-irradiance-*vs*-position plots for four-bar targets under specified phase distortions.

1.4 MTF and resolution

Resolution is a quantity without a standardized definition. Resolution can be defined as the separation, either in object-space angle or in image-plane distance, for which two discrete point targets can be easily discerned. Figure 1.16 illustrates the characteristic behavior of image-irradiance-*vs*-position plots in the spatial domain.

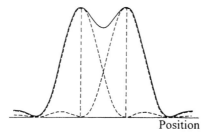

Figure 1.16 Resolution defined in the spatial domain.

Resolution can also be specified in the spatial-frequency domain as the inverse of this separation (cy/mm or cy/mrad) or as the limiting frequency at which the MTF falls below a particular threshold, the noise-equivalent modulation (NEM), seen in Fig. 1.17. We will discuss NEM more fully in Chapter 2.

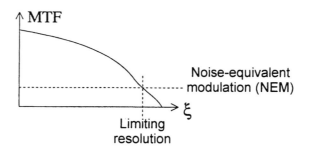

Figure 1.17 Resolution defined in the spatial-frequency domain.

Using either definition, resolution is a single-number performance specification and, as such, it is often seen as being more convenient to use than MTF (which is a function instead of a single number). However, MTF provides more complete performance information than is available from simply specifying resolution, including information about system performance over a range of spatial frequencies. As we can see on the left-hand side of Fig. 1.18, two systems may have identical limiting resolution, but different performances at lower frequencies. The system corresponding to the higher of the two curves would have the better image quality. The right-hand side of Fig. 1.18 shows that resolution alone can be a misleading performance criterion. The system that has the best resolution (limiting frequency) has poorer performance at the midrange frequencies. A decision about which system has better performance would require a specification of the spatial-frequency range of interest.

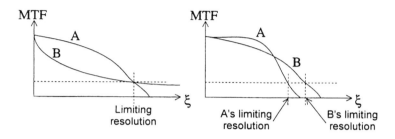

Figure 1.18   Resolution specifies only the limiting frequency, not the performance at other frequencies.

In general, the best overall image-quality performance is achieved by the imaging system that has the maximum area between the MTF and NEM curves over the frequency range of interest. This quantity, seen in Fig. 1.19, is called MTF area (MTFA).

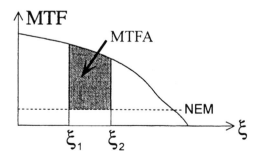

Figure 1.19   MTF area (MTFA) is the area between the MTF and NEM curves. Larger MTFA indicates better image quality.

1.5 Diffraction MTF

Because of the wave nature of light, an optical system with a finite-sized aperture can never form a point image. There is a characteristic minimum blur diameter formed, even when other image defects (aberrations) are absent. The best performance that the system is capable of is thus called "diffraction limited." The diameter of the diffraction blur is

$$d_{\text{diffraction}} = 2.44 \; \lambda \; (F/\#) \; . \tag{1.17}$$

The irradiance distribution in the diffraction image of a point source is

$$E(\rho) = |\; 2J_1(\pi\rho)/(\pi\rho) \;|^2 \; , \tag{1.18}$$

where ρ is the normalized radial distance from the center of the pattern,

$$\rho = r/(\lambda \; F/\#) \; . \tag{1.19}$$

A radial plot of Eq. (1.18) is shown in Fig. 1.20. To visualize the ring structure of the pattern, a two-dimensional plot is shown in Fig. 1.21. A two-dimensional integration of Eq. (1.18) shows that the blur diameter, $2.4 \; \lambda \; (F/\#)$ – the diameter to the first dark ring of the pattern – contains 84% of the power in the irradiance distribution.

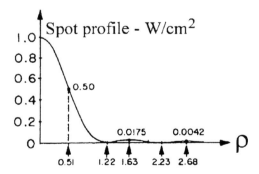

Figure 1.20 Radial plot of the diffracted irradiance distribution, Eq. (1.18) (adapted from Ref. 2).

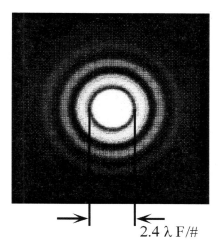

$2.4 \, \lambda \, F/\#$

Figure 1.21 Two-dimensional plot of the diffracted irradiance distribution.

In Eqs. (1.17) and (1.19) the parameter $F/\#$ is used as a scale factor that determines the physical size of the diffraction spot. Actually, three different expressions for $F/\#$ can be used for diffraction-limited spot size and diffraction-limited MTF calculations. As seen in Fig. 1.22 we can define a working $F/\#$ in either object space or in image space in terms of the lens aperture diameter D and either the object distance p or the image distance q:

$$(F/\#)_{\text{object-space}} = p/D \qquad (1.20)$$

or

$$(F/\#)_{\text{image-space}} = q/D \ . \qquad (1.21)$$

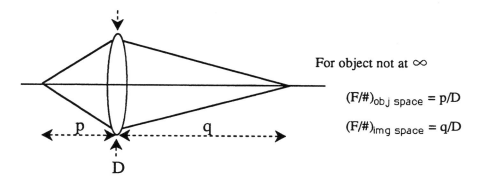

For object not at ∞

$(F/\#)_{\text{obj space}} = p/D$

$(F/\#)_{\text{img space}} = q/D$

Figure 1.22 Definition of working $F/\#$ for the finite-conjugate situation.

For the special case of an object at infinity (Fig. 1.23), the image distance q becomes the lens focal length f, and the image-space $F/\#$ becomes

$$(F/\#)_{\text{image-space}} = f/D \ . \tag{1.22}$$

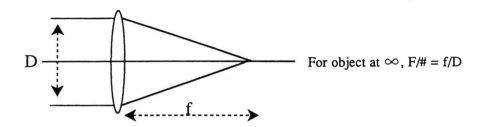

For object at ∞, F/# = f/D

Figure 1.23 Definition of image-space $F/\#$ for the object-at-infinity situation.

Using Eqs. (1.20), (1.21), or (1.22) we can calculate, either in object space or image space, the diffraction spot size from Eq. (1.17).

We can also consider the finite-sized spot caused by diffraction in terms of the MTF. Conceptually, a diffraction MTF can be calculated as the magnitude of the Fourier transform of diffraction impulse-response profile given in Eq. (1.18). We will perform the calculation under the assumption of an object at infinity. For an object not at infinity, we can use the appropriate $F/\#$ from Eqs. (1.20) or (1.21). Diffraction MTF is a wave-optics calculation for which the only variables (for a given aperture shape) are the aperture diameter D, wavelength λ, and focal length f. The MTF$_{\text{diffraction}}$ is the upper limit to the system's performance; the effects of optical aberrations are assumed to be negligible. Aberrations increase the spot size and thus contribute to a poorer MTF. The diffraction MTF is based on the overall limiting aperture of the system (the aperture stop). The diffraction effects are only calculated once per system and do not accumulate multiplicatively on an element-by-element basis.

1.5.1 Calculation of diffraction MTF

The diffraction OTF can be calculated as the normalized autocorrelation of the exit pupil of the system. We will show that this is consistent with the definition of Eqs. (1.6), (1.7), and (1.10), which state that the OTF is the Fourier transform of the impulse response. For the incoherent systems we consider, the impulse response $h(x,y)$ is the square of the two-dimensional Fourier transform of the diffracting aperture $p(x,y)$. The magnitude squared of the diffracted electric-field amplitude **E** in V/cm gives the irradiance profile of the impulse response in W/cm^2:

$$h_{\text{diffraction}}(x,y) = |\, \mathcal{FF}\, [\, p(x,y)\,]\,|^2 \quad . \tag{1.23}$$

From Eq. (1.23), we must implement a change of variables $\xi = x/\lambda f$ and $\eta = y/\lambda f$ to express the impulse response (which is a Fourier transform of the pupil function) in terms of image-plane spatial position. We then calculate the diffraction OTF in the usual way, as the Fourier transform of the impulse response $h(x,y)$:

$$\text{OTF}_{\text{diffraction}}(\xi,\eta) = \mathcal{FF}\, [\, h_{\text{diffraction}}(x,y)\,] = \mathcal{FF}\, \{\, |\, \mathcal{FF}\, [p(x,y)\,]\,|^2\, \} \quad . \tag{1.24}$$

Because of the absolute-value-squared operation, the two transform operations of Eq. (1.24) do not exactly undo each other – the diffraction OTF is the two-dimensional autocorrelation of diffracting aperture $p(x,y)$. The diffraction MTF is thus the magnitude of the (complex) diffraction OTF. As an example of this calculation, we take the simple case of a square aperture, seen in Fig. 1.24:

$$p(x,y) = \text{rect}(x/D, y/D) \quad . \tag{1.25}$$

The autocorrelation of the square is a triangle-shaped function,

$$\text{MTF}(\xi) = 1 - \xi/\xi_{\text{cutoff}} \, , \tag{1.26}$$

with cutoff frequency defined by

$$\xi_{\text{cutoff}} = 1/[\lambda\, (F/\#)] \quad . \tag{1.27}$$

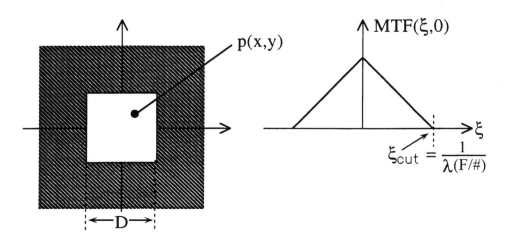

Figure 1.24 Terms for calculating the MTF of a square aperture.

For the case of a circular aperture of diameter D, the system has the same cutoff frequency, $\xi_{cutoff} = 1/(\lambda\ F/\#)$, but the MTF has a different functional form:

$$MTF(\xi/\xi_{cutoff}) = \frac{2}{\pi}\left\{\cos^{-1}(\xi/\xi_{cutoff}) - (\xi/\xi_{cutoff})\left[1 - (\xi/\xi_{cutoff})^2\right]^{1/2}\right\} \quad (1.28)$$

for $\xi < \xi_{cutoff}$ and

$$MTF(\xi/\xi_{cutoff}) = 0 \quad\quad\quad\quad\quad\quad\quad (1.29)$$

for $\xi > \xi_{cutoff}$. These diffraction-limited MTF curves are plotted in Fig. 1.25. The diffraction-limited MTF is an easy-to-calculate upper limit to performance; we need only λ and the $F/\#$ to compute it. An optical system cannot perform better than its diffraction-limited MTF—any aberrations will pull the MTF curve down. It is useful to compare the performance specifications of a given system to the diffraction-limited MTF curve to determine the feasibility of the proposed specifications, to decide how much headroom has been left for manufacturing tolerances, or to see what performance is possible within the context of a given choice of λ and $F/\#$.

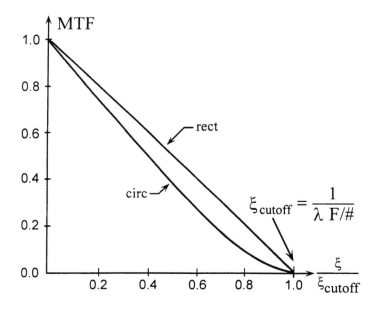

Figure 1.25 Universal curves for diffraction-limited MTFs, for incoherent systems with circular or rectangular aperture.

As an example of the calculations, let us consider the square-aperture system of Fig. 1.26(a) with an object at finite distance. Using the object-space or image-space $F/\#$ as appropriate, we can calculate ξ_{cutoff} in either the object plane or image plane:

$$\xi_{cutoff, obj} = \frac{1}{\lambda \, (F/\#)_{obj \, space}} = \frac{1}{\lambda \, (p/D)} = 666 \, cy/mm \qquad (1.30)$$

or

$$\xi_{cutoff, img} = \frac{1}{\lambda \, (F/\#)_{img \, space}} = \frac{1}{\lambda \, (q/D)} = 333 \, cy/mm \; . \qquad (1.31)$$

Because $p < q$, the image is magnified with respect to the object; hence a given feature in the object appears at a lower spatial frequency in the image, so the two frequencies in Eqs. (1.30) and (1.31) represent the same feature. The filtering caused by diffraction from the finite aperture is the same, whether considered in object space or image space. With the cutoff frequency in hand, we can answer questions such as for what image spatial frequency is the MTF 30%? We use Eq. (1.26) to find that 30% MTF is at 70% of the image-plane cutoff frequency, or 223 cy/mm. This calculation is for diffraction-limited performance. Aberrations will narrow the bandwidth of the system, so that the frequency at which the MTF is 30% will be lower than 223 cy/mm.

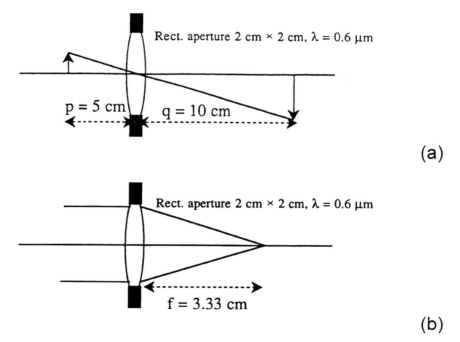

Figure 1.26   **(a) Rectangular-aperture finite-conjugate MTF example;**
(b) rectangular-aperture infinite-conjugate MTF example.

The next example shows the calculation for an object-at-infinity condition, with MTF obtained in object space as well as image space. We obtain the cutoff frequency in the image plane using Eq. (1.27):

$$\xi_{\text{cutoff, img space}} = 1/[\lambda \, (F/\#)] = D/\lambda f = 1000 \text{ cy/mm} . \qquad (1.32)$$

As in Fig. 1.26(a), a given feature in Fig. 1.26(b) experiences the same amount of filtering, whether expressed in the image plane or in object space. In the object space, we find the cutoff frequency is

$$\xi_{\text{cutoff, obj space}} = 1/(\lambda/D) = D/\lambda = 33.3 \text{ cy/mrad} . \qquad (1.33)$$

Let us verify that this angular spatial frequency corresponds to the same feature as that in Eq. (1.32). Referring to Fig. 1.27, we use the relationship between object-space angle θ and image-plane distance X,

$$X = \theta f . \qquad (1.34)$$

Inverting Eq. (1.34) to obtain the angular spatial frequency $1/\theta$:

$$1/\theta = (1/x) f = \xi f . \qquad (1.35)$$

Given that θ is in radians, if X and f have the same units, we can verify the correspondence between the frequencies in Eqs. (1.32) and (1.33):

$$(1/\theta) \, [\text{cy/mrad}] = \xi \, [\text{cy/mm}] \times f \, [\text{mm}] \times (0.001) \, [\text{rad/mrad}] . \qquad (1.36)$$

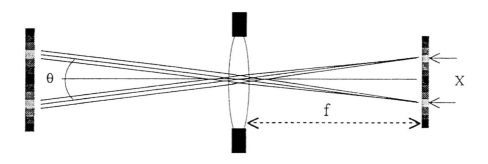

Figure 1.27 Relation between object-space angle and image-plane distance (adapted from Ref. 3).

It is also of interest to verify that the diffraction MTF curves in Fig. 1.25 are consistent with the results of the simple 84% encircled-power diffraction spot-size formula of 2.4 λ (*F/#*). In Fig. 1.28, we create a one-dimensional spatial frequency with adjacent lines and spaces. We approximate the diffraction spot as having 84% of its flux contained inside a circle of diameter 2.44 λ (*F/#*), and 16% in a circle of twice the diameter.

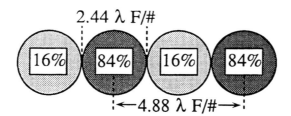

Figure 1.28 Verification of the formula for diffraction MTF.

The fundamental spatial frequency of the above pattern is

$$\xi = 1/[4.88 \ \lambda \ F/\#] = 0.21 \times [1/ \ (\lambda \ F/\#)] = 0.21 \ \xi_{\text{cutoff}} \qquad (1.37)$$

and the modulation depth at this frequency is

$$M = (84 - 16)/(84 + 16) = 68\% \ , \qquad (1.38)$$

in close agreement with Fig. 1.25 for a diffraction-limited circular aperture, at a frequency of $\xi = 0.21 \ \xi_{\text{cutoff}}$.

1.5.2 Diffraction MTFs for obscured systems

Many common optical systems, such as Cassegrain telescopes, have an obscured aperture. We can calculate the diffraction OTF of obscured-aperture systems according to Eq. (1.24), the autocorrelation of the pupil of the system. When there is an obscuration, a flux attenuation, which is proportional to the fractional blocked area of the pupil, occurs at all spatial frequencies. When the autocorrelation is calculated, we see a slight emphasis of the MTF at high frequencies corresponding to overlap of the clear part of the aperture in the shift-multiply-and-integrate operation of the autocorrelation. This emphasis has come at the expense of the overall flux transfer. In the usual definition of MTF to be unity at $\xi = 0$, this attenuation is normalized out when MTF is plotted.

In Fig. 1.29, the MTF curves for the obscured apertures exceed the diffraction-limited curve for no obscuration (curve A). This behavior is an artifact of the normalization, and there is actually no more image modulation at those frequencies than would exist in the case of an unobscured aperture. If we let the value of curve B at $\xi = 0$ be 0.84, the value of curve C at $\xi = 0$ be 0.75, and the value of curve D at $\xi = 0$ be 0.44, all of these MTF curves would be bounded as an upper limit by the diffraction-limited curve A for no obscuration.

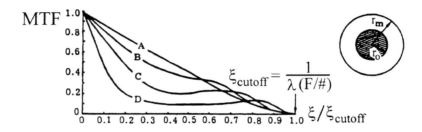

A:	no obscuration	0% area obscured;	100% area open
B:	$r_o = 0.25 \times r_m$	16% area obscured;	84% area open
C:	$r_o = 0.50 \times r_m$	25% area obscured;	75% area open
D:	$r_o = 0.75 \times r_m$	56% area obscured;	44% area open

Figure 1.29 Diffraction-limited MTF curves for obscured aperture systems (adapted from Ref. 4).

1.6 Geometrical MTF

We can calculate a geometrical OTF or MTF based on a ray-traced spot diagram for the system or based on geometrical-optics aberration formulae. Taking this spot diagram as $h(x,y)$ we can calculate a geometrical OTF in the manner of Eqs. (1.6) and (1.7) by Fourier transforming the impulse response. If the aberration blur is large compared to the diffraction spot, then we can use the geometrical OTF to represent the system performance. In the usual case of the aberration spot size being on the same order as the size of the diffraction spot, we can approximate the overall OTF by multiplying the geometrical and diffraction OTFs. Because of the convolution theorem of Eqs. (1.6) and (1.7), the multiplication of transfer functions is equivalent to convolving the diffraction spot with the aberration impulse response.[5,6] This approach does not provide the fine details of the diffraction image, but is accurate enough to provide a starting point for encircled-energy calculations.

1.6.1 Effect of defocus on MTF

In Fig. 1.30, we compare a diffraction-limited MTF to that for systems with increasing amounts of defocus. The amount of defocus is expressed in terms of optical path difference (OPD), the peak defocus at the edge of the aperture, in units of λ. In all cases, the diffraction-limited curve is the upper limit to the MTF. For small defocus the MTF curve is pulled down only slightly. For more defocus, there is a significant narrowing of the transfer function. In the case of a severely defocused system, we observe the phase-reversal phenomenon, the same as that seen in the spoke-target demonstration of Fig. 1.13.

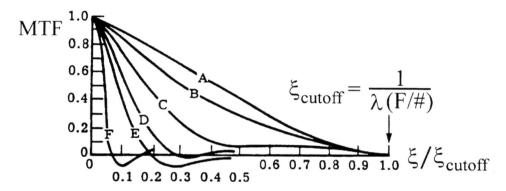

A -	focused: OPD = 0	D -	defocus: OPD = 3λ/4
B -	defocus: OPD = λ/4	E -	defocus: OPD = λ
C -	defocus: OPD = λ/2	F -	defocus: OPD = 5λ/4

Figure 1.30 Effect of defocus on the MTF of a diffraction-limited circular aperture system (adapted from Ref. 4).

1.6.2 Effect of other aberrations on MTF

Other aberrations affect MTF in a similar way as does defocus, pulling the MTF down, narrowing the curve, and lowering the cutoff frequency. Figure 1.31 illustrates that the value of the MTF at any spatial frequency is bounded by the diffraction-limited MTF curve:

$$\text{MTF}_{\text{w/aberr}}(\xi) \le \text{MTF}_{\text{diffraction}}(\xi) \; . \tag{1.39}$$

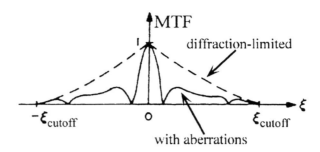

Figure 1.31   Effect of aberrations on MTF is to pull the transfer-function curve down (adapted from Ref. 2).

For plots of the effect of varying amounts of spherical aberration, coma, and astigmatism on MTF, see Williams and Becklund.[7]

1.6.3 MTF and Strehl ratio

A useful single-number performance specification is the Strehl ratio, s, defined as the on-axis irradiance produced by the actual system, divided by on-axis irradiance that would be formed by a diffraction-limited system of the same $F/\#$:

$$s \equiv \frac{h_{\text{actual}}(x = 0, y = 0)}{h_{\text{diffraction}}(x = 0, y = 0)} \quad . \tag{1.40}$$

A Strehl ratio in excess of 0.8 indicates excellent image quality ($\approx \lambda/4$ OPD). We can obtain an alternate interpretation of the Strehl ratio using the central-ordinate theorem for Fourier transforms, which states that the area under a function in the transform domain is equal to the on-axis value of the function in the spatial domain

$$f(x = 0, y = 0) = \iint F(\xi, \eta)\, d\xi\, d\eta \quad . \tag{1.41}$$

Thus, we can also express the Strehl ratio as the area under the actual OTF curve divided by the area under the diffraction-limited OTF curve:

$$s = \frac{\iint OTF_{\text{actual}}(\xi, \eta)}{\iint OTF_{\text{diffraction}}(\xi, \eta)} \quad . \tag{1.42}$$

Aberration effects such as those seen in Fig. 1.31 can then be interpreted directly in terms of the decrease in area under the MTF curve. We can explore the relation between Strehl ratio and MTF by comparing the $\lambda/4$-of-defocus condition in terms of the impulse response and the MTF. For this small amount of defocus, the transfer function is purely real and OTF = MTF. From Fig. 1.32, we see that the aberration moves a bit of power from the center of the impulse response and distributes it into the ring structure. The on-axis value of the impulse response is reduced, resulting in a corresponding reduction in the area under the MTF curve. It is somewhat easier to see the effect of this small amount of defocus on the MTF, particularly if the impulse response were normalized to 1 at the center.

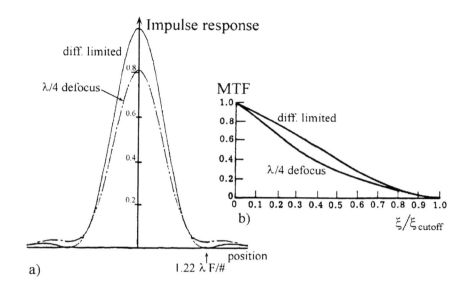

Figure 1.32 Comparison of (a) Strehl ratio (adapted from Ref. 8) and (b) MTF (adapted from Ref. 4) for a quarter wavelength of defocus.

References

1. M. Beran and G. Parrent, *Theory of Partial Coherence*, Prentice-Hall, Englewood Cliffs, NJ (1964).

2. J. Gaskill, *Linear Systems, Fourier Transforms, and Optics*, Wiley, New York (1978), referenced figures reprinted by permission of John Wiley and Sons, Inc.

3.   E. Dereniak and G. D. Boreman, *Infrared Detectors and Systems*, Wiley, New York (1996), referenced figures reprinted by permission of John Wiley and Sons, Inc.

4.   W. Wolfe and G. Zissis, eds., *The Infrared Handbook*, ERIM/IRIA, Ann Arbor (1978), referenced figures reprinted by permission.

5.   E. Linfoot, "Convoluted spot diagrams and the quality evaluation of photographic images," *Optica Acta* 9, p. 81 (1962).

6.   K. Miyamoto, "Wave optics and geometrical optics in optical design," in *Progress in Optics, Vol. 1,* E. Wolf, ed., North-Holland, Amsterdam (1961).

7.   C. Williams and O. Becklund, *Introduction to the Optical Transfer Function*, Wiley, New York (1989).

8.   J. Wyant, class notes, University of Arizona; also see *Applied Optics and Optical Engineering Vol. XI*, R. Shannon and J. Wyant, eds., Academic Press, Orlando (1992).

Further reading

L. R. Baker, *Selected Papers on Optical Transfer Function: Foundation and Theory*, SPIE Milestone Series, Vol. MS 59, 1992.

T. L. Williams, *The Optical Transfer Function of Imaging Systems*, Inst. of Physics Press, Bristol (1999).

CHAPTER 2
MTF IN ELECTRO-OPTICAL SYSTEMS

In Chapter 1 we applied a transfer-function-based analysis to describe image quality in classical optical systems--that is, systems with glass components only. In this chapter we will examine the MTF of electro-optical systems--that is, systems that use a combination of optics, scanners, detectors, electronics, signal processors, and displays.

To apply MTF concepts in the analysis of electro-optical systems, we must modify our assumptions about MTF. Electro-optical systems typically include detectors or detector arrays for which the size of the detectors and the spatial sampling interval are both finite. We will therefore develop an expression for the MTF impact of irradiance averaging over the finite sensor size. Also, because of the shift-variant nature of the impulse response for sampled-data systems, we will develop the concept of an average impulse response obtained over a statistical ensemble of source positions. Perhaps most important, one cannot amplify an arbitrarily small signal and obtain a useful result. The classical MTF theory presented in Chapter 1 does not account for the effects of noise. Noise is inherent in any system with electronics. We will demonstrate how to broaden the MTF concept to include this important issue.

2.1 Detector footprint MTF

We often think about the object as being imaged onto the detectors, but it is also useful to consider where the detectors are imaged. The footprint of a particular detector, called the instantaneous field of view (IFOV), is the projection of that detector into object space. We consider a scanned imaging system in Fig. 2.1, and a staring focal-plane-array (FPA) imaging system in Fig. 2.2. In each case, the flux falling onto an individual detector produces a single output. Inherent in the finite size of the detector elements is some spatial averaging of the image irradiance. For the configurations shown, we have two closely spaced point sources in the object plane that fall within one detector footprint. The signal output from the sensor will not distinguish the fact that there are two sources. Our first task is to quantify the spatial-frequency filtering inherent in an imaging system with finite-sized detectors.

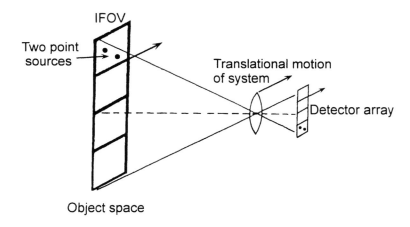

Figure 2.1 Scanned imaging system.

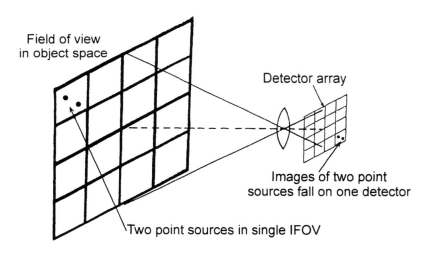

Figure 2.2 Staring focal-plane-array imaging system.

A square detector of size $w \times w$ performs spatial averaging of the scene irradiance that falls on it. When we analyze the situation in one dimension, we find the integration of the scene irradiance $f(x)$ over the detector surface is equivalent to a convolution of $f(x)$ and the rect function[1] that describes the detector responsivity:

$$g(x) = \int_{-w/2}^{w/2} f(x)\,dx = f(x) * \mathrm{rect}(x/w) \ . \tag{2.1}$$

By the convolution theorem, Eq. (2.1) is equivalent to filtering in the frequency domain by a transfer function

$$\mathrm{MTF}_{\mathrm{footprint}}(\xi) = \left| \mathrm{sinc}(\xi w) \right| = \left| \frac{\sin(\pi \xi w)}{\pi \xi w} \right| . \qquad (2.2)$$

Equation (2.2) shows us that the smaller the sensor photosite dimension is, the broader will be the transfer function. Equation (2.2) is a fundamental MTF component for any imaging system with detectors. In any given situation, the detector footprint may or may not be the main limitation to image quality, but its contribution to a product such as Eq. (1.7) is always present. Equation (2.2) is plotted in Fig. (2.3) and we see that the sinc-function MTF has its first zero at $\xi = 1/w$. It is instructive to consider the following plausibility argument to justify the fact that the footprint MTF = 0 at $\xi = 1/w$.

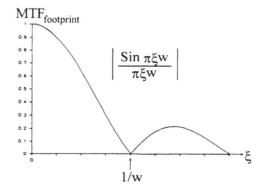

Figure 2.3 Sinc-function MTF for detector of width _w_.

Consider the configuration of Fig. 2.4, representing spatial averaging of an input irradiance waveform by sensors of a given dimension w. The individual sensors may represent either different positions for a scanning sensor or discrete locations in a focal-plane array. We will consider the effect of spatial sampling in a later section. Here we consider exclusively the effect of the finite size of the photosensitive regions of the sensors. We see that at low spatial frequencies there is almost no reduction in modulation of the image irradiance waveform arising from spatial averaging over the surfaces of the photosites. As the spatial frequency increases, the finite size of the detectors becomes more significant. The averaging leads to a decrease in the maximum values and an increase in the minimum values of the image waveform – a decrease in the modulation depth. For the spatial frequency $\xi = 1/w$, one period of the irradiance waveform just fits onto each detector. Regardless of the position of the input waveform with respect to the photosite boundaries, each sensor will collect exactly the same power (integrated irradiance) level. The MTF is zero at $\xi = 1/w$ because each sensor reads the same level and there is no modulation depth in the resulting output waveform.

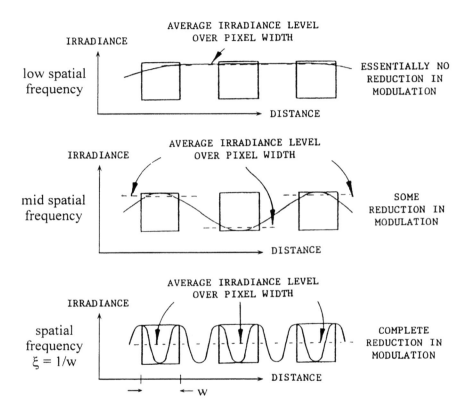

Figure 2.4 At a frequency of $\xi = 1/w$ the modulation depth goes to zero.

Extending our analysis to two dimensions, we consider the simple case of a rectangular detector with different widths along the x and y directions:

$$h_{\text{footprint}}(x) = \text{rect}(x/w_x, y/w_y) = \text{rect}(x/w_x)\, \text{rect}(y/w_y) . \qquad (2.3)$$

By Fourier transformation, we obtain the OTF, which is a two-dimensional sinc function:

$$\text{OTF}_{\text{footprint}}(\xi, \eta) = \text{sinc}(\xi w_x, \eta w_y) \qquad (2.4)$$

and

$$\text{MTF}_{\text{footprint}}(\xi, \eta) = \left| \frac{\sin(\pi\xi w_x)}{\pi\xi w_x} \right| \left| \frac{\sin(\pi\eta w_y)}{\pi\eta w_y} \right|. \qquad (2.5)$$

The impulse response in Eq. (2.3) is separable; that is, $h_{\text{footprint}}(x,y)$ is simply a function of x multiplied by a function of y. The simplicity of the separable case is that both $h(x,y)$ and $H(\xi,\eta)$ are products of two one-dimensional functions, with the x and y dependences completely separated. Occasionally a situation may arise in which the detector responsivity function is not separable.[2,3] In that case, we can no longer write the MTF as the product of two one-dimensional MTFs. The MTF along the ξ and η spatial frequencies is affected by both x and y profiles of detector footprint. For example, the MTF along the ξ direction is not simply the Fourier transform of the x profile of the footprint but is

$$\text{MTF}_{\text{footprint}}(\xi) = \left| H_{\text{footprint}}(\xi, \eta = 0) \right| \neq \left| \mathcal{F}[h_{\text{footprint}}(x, y = 0)] \right|. \qquad (2.6)$$

Analysis in these situations requires a two-dimensional Fourier transform of the detector footprint. The transfer function can then be evaluated along the ξ or η axis, or along any other desired direction.

2.2 Sampling

Sampling is a necessary part of the data-acquisition process in any electro-optical system. We will sample at spatial intervals $\Delta x \equiv x_{\text{samp}}$. The number of samples that we take and hence the spatial sampling rate are constrained by digital storage or processing capacity or by the number of detectors in the system. Sampling is often insufficient because of cost constraints: we typically want to cover as wide a field of view as possible while using a fewer-than-optimum number of detectors and samples. The process of spatial sampling has two main effects on image quality: aliasing and the sampling MTF.

2.2.1 Aliasing

Aliasing is an image artifact that occurs when we insufficiently sample a waveform. We will consider an image waveform that has been decomposed into its constituent sinusoids: a waveform of spatial frequency ξ. If we choose a sampling interval sufficiently fine to locate the peaks and valleys of the sine wave, then we can reconstruct that particular frequency component unambiguously from its sampled values. The two-samples-per-cycle minimum sampling rate seen in Fig. 2.5 corresponds to the Nyquist condition

$$\Delta x \equiv x_{\text{samp}} < 1/(2\xi) , \tag{2.7}$$

with the inequality required to ensure that all values of the waveform will be sampled. Samples acquired at an interval equal to $1/(2\xi)$ could conceivably all be at the same level (the 50%-amplitude point of the sinusoid) so, from a practical point of view, the Nyquist condition is represented by the inequality of Eq. (2.7).

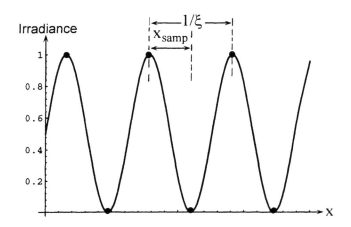

Figure 2.5 Nyquist sampling condition of two samples per cycle.

If the sampling is less frequent ($x_{\text{samp}} > 1/(2\xi)$) than required by the Nyquist condition, then we will see the samples as representing a lower-frequency sinewave (Fig. 2.6). Even though both sine waves shown are consistent with the samples, we will perceive the low-frequency waveform when looking at the sampled values. This image artifact, where samples of a high-frequency waveform appear to represent a low-frequency waveform, is an example of aliasing. Aliasing is symmetric about the Nyquist frequency, which means that the amount by which a waveform exceeds Nyquist is the amount at which we perceive it to be below Nyquist. So the frequency transformation of

$$(\xi_{\text{Nyquist}} + \Delta\xi) \rightarrow (\xi_{\text{Nyquist}} - \Delta\xi) \tag{2.8}$$

takes place between the input waveform and the aliased image data. Figure 2.7 shows a pictorial example of aliasing for the case of a radial grating, for which the spatial frequency of the target increases toward the center. With an insufficient spatial-sampling rate we see that the high frequencies near the center are aliased into lower spatial frequencies.

Irradiance

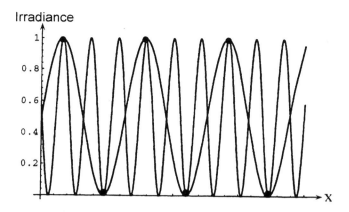

Figure 2.6 Aliasing of a high-frequency waveform to a lower spatial frequency by insufficient sampling.

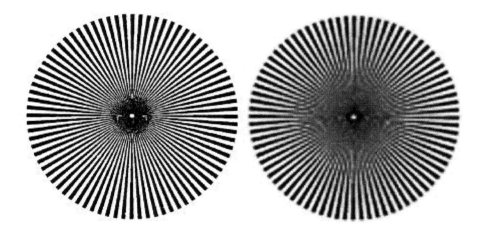

Figure 2.7 Example of aliasing in a radial grating. The right-hand image has been sampled at a larger sampling interval, and hence cannot adequately represent the high frequencies near the center of the pattern.

Figure 2.8 is a bar-target pattern that shows aliasing artifacts. We acquired the image on the left-hand side with a higher sampling frequency; thus the bars have equal lines and spaces and are of equal density. In the right-hand image, even though the fundamental spatial frequencies (inverse of the spacings) of all of the bars are below the Nyquist frequency and are hence adequately sampled, the highest frequencies in the harmonics of the bar patterns are aliased. The fact that not all of the bars in a given three-bar pattern are of the same width or density in the undersampled image is evidence of aliasing at high frequencies.

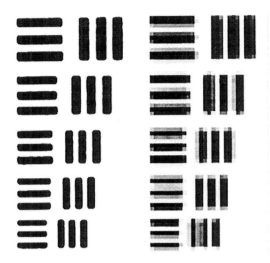

Figure 2.8 Bar-target pattern showing aliasing artifacts. The right-hand image has been sampled using a wider-spaced sampling interval, and hence cannot adequately represent the high-frequency harmonics of the bars.

After we have sampled the waveform, we cannot remove aliasing artifacts by filtering because, by Eq. (2.8), the aliased components have been lowered, to fall within the main spatial-frequency passband of the system. For us to remove aliasing artifacts at this point requires the attenuation of broad spatial-frequency ranges of the image data. We can avoid aliasing in the first place by prefiltering the image, that is, bandlimiting it before the sampling occurs. The ideal anti-aliasing filter would pass at unit amplitude all frequency components for which $\xi < \xi_{Nyquist}$ and attenuate completely all components for which $\xi > \xi_{Nyquist}$. The problem is that neither the detector MTF (a sinc function) nor the optics MTF (bounded by an autocorrelation function) follows the form of the desired anti-aliasing filter seen in Fig. 2.9.

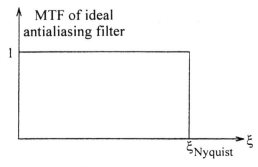

Figure 2.9 Ideal anti-aliasing filter.

An abrupt filter shape such as the one in Fig. 2.9 can easily be implemented in the electronics subsystem. However, at that stage the image irradiance has already been sampled by the sensors, so the electrical filter cannot effectively serve an anti-aliasing function. The sinc function corresponding to a rectangular detector can serve as an anti-aliasing filter. It is most effective in the situation where the image data are more finely sampled than once per detector element because then the aliasing frequency is sufficiently high that the detector MTF is relatively low at Nyquist. For contiguous detectors (one sample per detector element) the sampling distance is equal to the detector width: $x_{samp} = w$. In this case the first zero of the detector MTF occurs at $\xi = 1/w$, which is twice the Nyquist frequency of $1/(2x_{samp})$. Thus, for the contiguous detector case, the MTF is relatively high at the onset of aliasing. Finer sampling in the spatial domain leads to a higher Nyquist frequency and a consequently lower MTF at Nyquist. This situation can be implemented in scanned systems, which are addressed later in this chapter.

The optics MTF offers some flexibility as an anti-aliasing filter but, because it is bounded by the autocorrelation function of the aperture, it does not allow for the abrupt-cutoff behavior of Fig. 2.9. By choice of λ and $F/\#$ and by the addition of aberrations such as defocus, we can control the cutoff frequency of the optics MTF. But the approximately linear falloff with frequency forces us to trade off reduced MTF at frequencies less than $\xi_{Nyquist}$ against the amount of residual aliasing. Compared to the ideal filter seen in Fig. 2.9, use of the linear-falloff diffraction MTF as in Eq. (1.23) and Fig. 1.24 as an anti-aliasing filter requires setting the cutoff so that MTF($\xi \geq \xi_{Nyquist}$) = 0. As seen in Fig. 2.10, this setting results in a loss of considerable area under the MTF curve at frequencies below Nyquist. If we set the cutoff frequency higher, additional modulation depth for $\xi < \xi_{Nyquist}$ is preserved at the expense of nonzero MTF above Nyquist (leading to some aliasing artifacts). The choice of a higher cutoff frequency for the linear MTF function preserves more modulation below Nyquist, but results in higher visibility of aliasing artifacts.

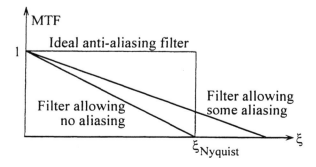

Figure 2.10 Tradeoff of MTF below Nyquist with amount of residual aliasing for a filter corresponding to diffraction-limited optics MTF.

Birefringent filters that are sensitive to the polarization state of the incident radiation can be configured to perform an anti-aliasing function,[4] although still without the ideal abrupt-cutoff MTF shown in Fig. 2.9. A filter of the type shown in Fig. 2.11 is particularly useful in color focal-plane arrays, where different spectral filters (red, blue, green) are placed on adjacent photosites. Because most visual information is received in the green portion of the spectrum the sampling interval for the red- and blue-filtered sensors is wider than for the green-filtered sensors. If we consider each color separately, we find a situation equivalent to the sparse-array configuration seen in Fig. 2.11, where the active photosites for a given color are shown shaded. The function of the filter is to split an incident ray into two components. A single point in object space maps to two points in image space, with a spacing equal to one-half the sensor-to-sensor distance. The impulse response of the filter is two delta functions:

$$h_{\text{filter}}(x) = \tfrac{1}{2}[\delta(x) + \delta(x + x_{\text{samp}}/2)] \ . \tag{2.9}$$

The corresponding filter transfer function can be found by Fourier transformation as

$$\text{MTF}_{\text{filter}}(\xi) = |\cos[2\pi(x_{\text{samp}}/2)\xi]| \ , \tag{2.10}$$

which has its first zero at $1/(2x_{\text{samp}})$, which is ξ_{Nyquist}. The filter cuts off at the right location and falls to zero faster than either the linear-falloff optics MTF, or the contiguous-detector sinc function MTF with cutoff at $2\xi_{\text{Nyquist}}$.

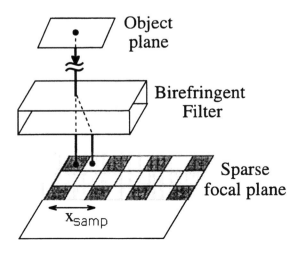

Figure 2.11 Mechanism of a birefringent anti-aliasing filter.

2.2.2 Sampling MTF

We can easily see that a sampled-imaging system is not shift invariant. Consider the focal-plane-array (FPA) imager seen in Fig. 2.12. The position of the image-irradiance function with respect to the sampling sites will affect the final image data. If the image is aligned so that most image irradiance falls completely on one single column of the imager, then a high-level signal is produced that is spatially compact. If the image irradiance function is moved slightly so that it falls on two adjacent columns, the flux from the source is split in two and a lower-level broader signal is read. If we measure the MTF for such a sampled system, the Fourier transform of the spatial-domain image will depend on the alignment of the target and the sampling sites, with best alignment giving the broadest MTF.

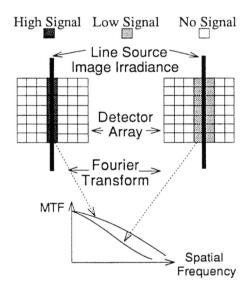

Figure 2.12   A sampled image-forming system is not spatially invariant (adapted from Ref. 5).

This shift variance violates one of the main assumptions required for a convolutional analysis of the image-forming process. To preserve the convenience of a transfer-function approach, we need to find a way to generalize the concept of impulse response to define a shift invariant quantity. Following Park et al.,[6] we define a spatially averaged impulse response and a corresponding MTF component that are inherent in the sampling process itself by assuming that the scene being imaged is randomly positioned with respect to the sampling sites. This random alignment corresponds to the situation where a natural scene is imaged with an ensemble of individual alignments. For a two-dimensional rectangular sampling grid, the sampling impulse response is a rectangle function whose widths are equal to the sampling intervals in each direction:

$$h_{sampling}(x,y) = \text{rect}(x/x_{samp}, y/y_{samp}) \ . \tag{2.11}$$

We see from Eq. (2.11) that wider-spaced sampling produces an image with poorer image quality. An average sampling MTF can be defined as the magnitude of the Fourier transform of $h_{sampling}(x,y)$

$$\text{MTF}_{sampling} = |\ \mathcal{F}\ [\text{rect}(x/x_{samp}, y/y_{samp})]\ | \ , \tag{2.12}$$

which yields a sinc-function sampling MTF of

$$\text{MTF}_{sampling} = \left| \text{sinc}(\xi x_{samp}, \eta y_{samp}) \right| = \frac{\sin(\pi \xi x_{samp})}{\pi \xi x_{samp}} \frac{\sin(\pi \eta y_{samp})}{\pi \eta y_{samp}} \ . \tag{2.13}$$

The sampling MTF is equivalent to the average of the MTFs that would be realized for an ensemble of source locations, uniformly distributed with respect to the sampling sites. As Fig. 2.12 demonstrates, when the alignment is optimum the MTF is broad, but for other source positions the MTF is narrower. The sampling MTF is the average over all possible MTFs. Thus defined, the average is a shift-invariant quantity, so that we can proceed with a usual transfer-function-based analysis. The sampling MTF of Eq. (2.13) is a component that multiplies the other MTF components for the system.

However, this sampling MTF does not contribute in a MTF-measurement setup where the test target is aligned with the sampling sites because the central assumption in its derivation is the random position of any image feature with respect to the sampling sites. In typical MTF test procedures, the position of the test target is adjusted to yield the best output signal (most compact output, best appearance of bar-target images). In the typical test-setup case, the sampling MTF equals unity except where random-noise test targets[7] that explicitly include the sampling MTF in the measurement are used. Because typical test procedures preclude the sampling MTF from contributing to the measurements, the sampling MTF is often forgotten in a systems analysis. However, when the scene being imaged has no net alignment with respect to the sampling sites, the sampling MTF will contribute in practice and should thus be included in the system-performance modeling.

The combined MTF of the optics, detector-footprint, and sampling can be much less than initially expected, especially considering two common misconceptions that neglect the detector and sampling MTFs. The first error is to assume that if the optics blur-spot size is matched to the detector size then there is no additional image-quality degradation from the finite detector size. We can see from Eq. (1.9) that the optics and detector MTFs multiply, and hence both terms contribute. Also, it is quite common to forget the sampling MTF, which misses another multiplicative sinc contribution. As an illustration of the case where the optics blur-spot size is matched to the detector size in a contiguous FPA, we assume that each impulse response can be modeled in a one-dimensional analysis as a rect function of width w_x. Each of the three MTF terms is a sinc function and the product is proportional to $\text{sinc}^3(\xi w_x)$, as seen in Fig. 2.13. The difference between the expected $\text{sinc}(\xi w_x)$ and the actual $\text{sinc}^3(\xi w_x)$ at midrange spatial frequencies can be 30% or more in terms of absolute MTF and, at high frequencies, the MTF can go from an expected value of 20% to essentially zero, with the inclusion of the two additional sinc-function contributions.

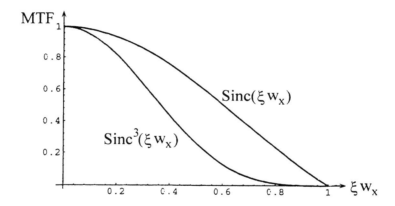

Figure 2.13 MTF contributions multiply for the detector footprint, optics blur spot, and sampling.

We now compare several sampled-image situations where we consider the overall MTF as the product of sampling MTF and footprint MTF. In these examples we consider only the MTF inherent in the averaging-and-sampling process arising from finite-sized detectors that have a finite center-to-center spacing. In actual application, other MTF contributions such as those arising in the optics or electronics subsystems would multiply these results. For simplicity, we analyze only the x and y sampling directions. Recall that the Nyquist frequency is the inverse of twice the sampling interval in each direction. In each of the cases considered, we keep the sensor dimension constant at w and investigate the aggregate MTF as we vary the sampling situation. We do not account for the finite dimension of the array as a whole in any of the examples.

First, we consider the sparse FPA shown in Fig. 2.14. In this case, we take the sampling interval to be twice the detector width. The x and y directions are identical. Because the Nyquist frequency is rather low at $\xi = 0.25/w$, the MTF is high at the aliasing frequency, which means that aliasing artifacts such as those seen in Figs. (2.7) and (2.8) are visible with high contrast. In addition, the large sampling interval places the first zero of the sampling MTF at $\xi = 0.5/w$, narrowing the MTF product considerably compared to the detector MTF, which has a first zero at $\xi = 1/w$.

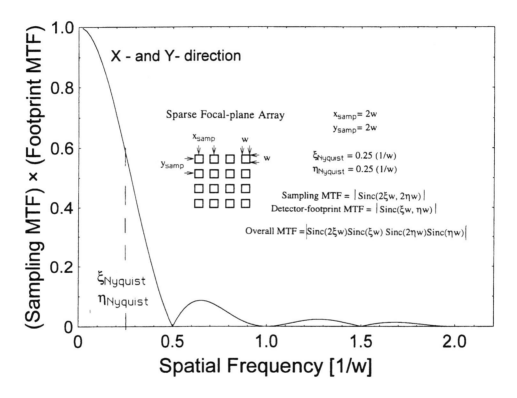

Figure 2.14 MTF for a sparse focal-plane array.

In Fig. 2.15, the sensor size remains the same as that in Fig. 2.14, but now the detectors are contiguous with a sampling interval twice the detector width. The Nyquist frequency is thus raised to $\xi = 0.5/w$, which has two related effects. First, because the aliasing frequency is higher, the MTF is lower at the aliasing frequency, so that the artifacts are not as visible. Also, the usable bandwidth of the system, from dc to the onset of aliasing, has been increased by the finer sampling. The sinc-function MTFs for the detectors and for the sampling are identical, with a first zero at $\xi = 1/w$ for each. Their product is a sinc-squared function, which has considerably higher MTF than did the MTF of the sparse array seen in Fig. 2.14.

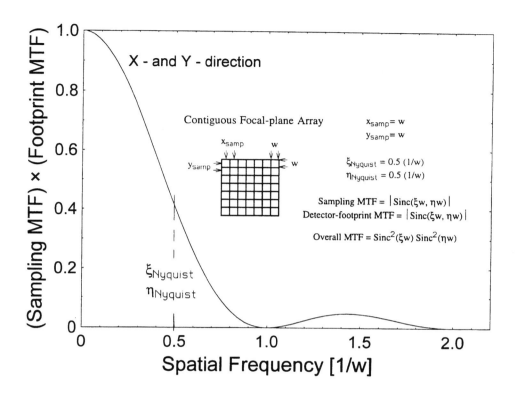

Figure 2.15 MTF for a contiguous focal-plane array.

Sparse FPAs can be made to have MTFs equivalent to those of contiguous FPAs if we employ microscanning (also called microdither) techniques, which include the use of piezoelectric actuators or liquid-crystal beam steerers to repetitively move the FPAs or the line of sight of the image-delivery optics.[8,9] We thus obtain successive frames of displaced image samples in what would otherwise be dead space between detectors (Fig. 2.16). Finer sampling yields better sampling MTF, along with higher Nyquist frequencies. This approach can also be applied to FPAs that are contiguous to begin with, to further decrease the sampling interval. The fact that microscanning produces better pictures is intuitive proof of the existence of a sampling MTF contribution, because the sensor size and the optics MTF are not changed by microscanning. The single MTF component that is improved by the microscan technique is the sampling MTF. The drawback to microscanning, from a systems viewpoint, is that the frame takes longer to acquire for a given integration time. Alternatively, if we keep the frame rate constant, the integration time decreases, which can have a negative impact on signal-to-noise ratio. There is also an additional data-processing complexity of interlacing the frames together.

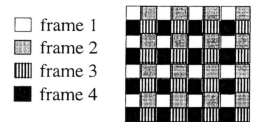

□ frame 1
▦ frame 2
▥ frame 3
■ frame 4

Figure 2.16   Microscan technique for increasing the sampling MTF. Detector IFOVs are moved on successive frames and interlaced together.

    Referring to Fig. 2.17, we explore the sampling situation with a linear array of scanned detectors. The designer decides the sampling interval for a continuously scanned detector; with choices limited only by the desired acquisition speed and the total number of data samples. Typical practice is to sample the analog signal from the detector at time intervals equivalent to two samples per detector width, in other words, twice per dwell. Finer sampling is certainly possible, but we obtain the maximum increase in image quality by going from one to two samples per dwell. Beyond that spatial sampling rate, there is a diminishing return in terms of image quality. Let us see why two samples per dwell has been so popular an operating point. With a sampling interval of $w/2$, the x-direction Nyquist frequency has been increased to $\xi = 1/w$. This higher aliasing frequency is beneficial in itself because the usable bandwidth has been increased, but the other factor is that now the MTF of the detector footprint goes through its first zero at the Nyquist frequency. Because the transfer function is zero at Nyquist, the image artifacts arising from aliasing are naturally suppressed. A final advantage is that the x-direction MTF has been increased by the broader sampling-MTF sinc function, which has its first zero at $\xi = 2/w$. Because the detectors are contiguous in the y direction, the aliasing frequency is lower at $\eta = 0.5/w$ and the MTF as a function of η is just the sinc-squared function seen in the analysis of the contiguous FPA in Fig. 2.15.
    In Fig. 2.18 we extend this analysis to a pair of staggered linear arrays offset by one-half the detector-to-detector spacing. Once again, we must perform additional data processing to interlace the information gathered from both sensor arrays into a high-resolution image. The advantage we gain is that there is a twice-per-dwell sampling in both the x and y directions, with wider η-direction MTF, higher η-direction aliasing frequency, and additional suppression of aliasing artifacts. Although it is possible to interlace more than two arrays, the greatest image-quality benefit is gained in going from a single array to two.

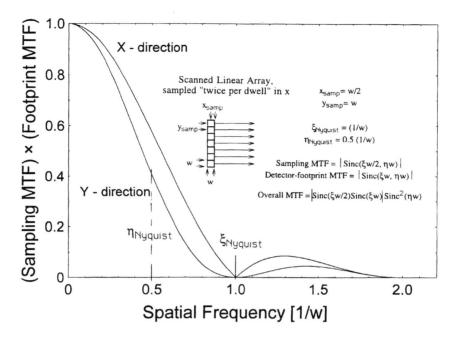

Figure 2.17 MTF for a scanned linear array of sensors.

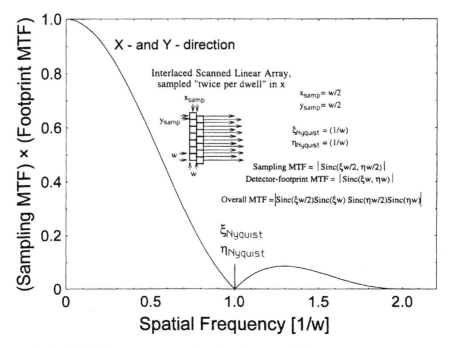

Figure 2.18 MTF for a staggered pair of scanned linear arrays.

   As a final example of this type of analysis, we consider a fiber array
(Fig. 2.19), which can transmit an image-irradiance distribution over the length
of the fibers in the bundle.[10,11] If we assume that the arrangement of the fibers is
preserved between the input and output faces (a so-called coherent array), we
can describe the MTF as the product of the MTF of the fiber footprint (because
there is no spatial resolution within an individual fiber, only a spatial averaging)
and the sampling MTF (which depends on the details of the arrangement of the
fibers). In this example, we have a hexagonal packed array, with an x-sampling
interval of $D/2$ and a y-sampling interval of $\sqrt{3}\,D/2$. Other spatial arrangements
of the fibers are possible. Once the center-to-center spacing of the fibers in each
direction is determined, the sampling MTF along ξ and η can be found from Eq.
(2.13), assuming that any scene irradiance distribution is randomly positioned
with respect to the fibers. The Fourier transform of the fiber footprint yields the
Bessel-function MTF shown in Fig. 2.19.

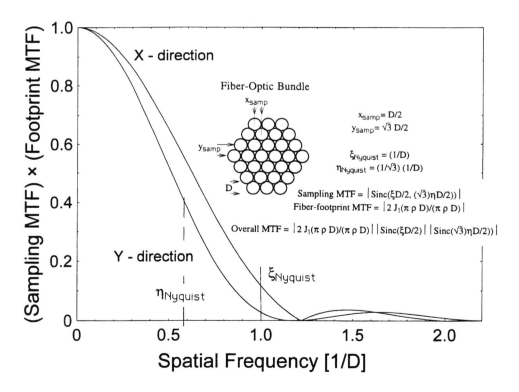

Figure 2.19 MTF for a fiber-bundle array.

So far, we have considered the nearest-neighbor sampling distances along the x and y directions. If we extend this sampling to any direction in the x-y plane, we can extend our applications to any sampled-image system where directions other than x and y are important for image formation, such as hexagonal focal-plane arrays, fiber bundles, and laser printers. Once the two-dimensional sampling MTF is in hand, we multiply it by the two-dimensional Fourier transform of pixel footprint, to yield the overall sampling-and-averaging array MTF.[12,13] A one-dimensional sinc-function sampling MTF along the lines of Eq. (2.13) applies to the spacing between the nearest neighbors in any direction because the distance between samples in any direction can be modeled as a rect-function impulse response (assuming random position of the scene with respect to the sampling sites). The width of the rect depends on the particular direction θ in which the next-nearest neighbor is encountered:

$$h_{\text{sampling}}[x(\theta)] = \text{rect}\,[x/x_{\text{samp}}(\theta)] \qquad (2.14)$$

and

$$\text{MTF}_{\text{sampling}}(\xi_\theta) = |\,\mathcal{F}\,\{\text{rect}\,[x/x_{\text{samp}}(\theta)]\} = |\,\text{Sinc}[\xi_\theta x_{\text{samp}}(\theta)]\,|\,, \qquad (2.15)$$

where ξ_θ is understood to be a one-dimensional spatial frequency along the θ direction. Directions with wider spaced next-nearest neighbors will have poorer image quality. For a finite-dimension sampling array, a nearest neighbor does not exist in some directions at all, and thus the MTF is necessarily zero in that direction (because the image data of that particular spatial frequency cannot be reconstructed from the samples). We find that the MTF of Eq. (2.15) is thus a discontinuous function of angle (Fig. 2.20).

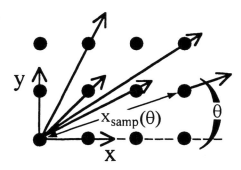

Figure 2.20 Nearest-neighbor distance, and hence sampling MTF, is a discontinuous function of angle θ.

In Fig. 2.21, we consider a pictorial example of aliasing, sampling MTF, and detector-footprint MTF. We look at three versions of the same scene, with a fixed field of view. In each case, the detectors are contiguous ($x_{samp} = w_x$). Version a) is a 512×512 image. It has good image quality and appears to be spatially continuous. Version b) is the same image, sampled at 128×128. Aliasing artifacts begin to be visible, particularly in the folds of the shirt, because of the lower Nyquist frequency. We begin to see some blurring effects because of the combined impact of the poorer sampling and detector-footprint MTFs. Version c) is the same image, sampled at 64×64. Extensive aliasing artifacts are seen, visible as low-frequency banding, again particularly in the folds of the shirt, but also in the edges of the head and the shirt. The image quality is dominated by the dimensions of the detector and the sampling interval.

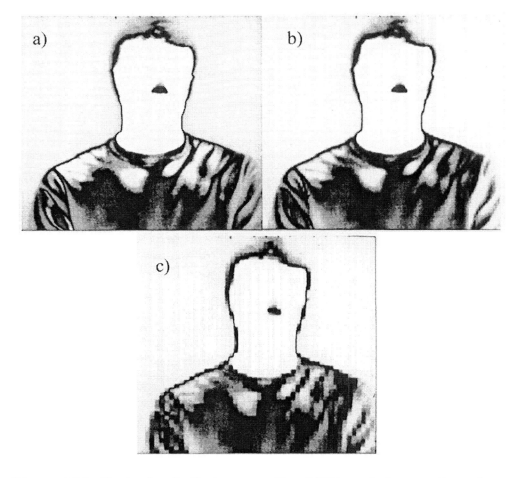

Figure 2.21   Example of aliasing, sampling MTF, and detector-footprint MTF.

2.3 Crosstalk

Crosstalk arises when the signal of a particular detector contributes to or induces a spurious signal on its neighbor. Origins of crosstalk include charge-transfer inefficiency in charge-coupled devices, photogenerated carrier diffusion,[14] or coupling of close capacitors. As Fig. 2.22 shows, crosstalk can be approximately modeled with an impulse response of a Gaussian or negative-exponential form. We typically do not have a large number of sample points, because only the nearest few channels have an appreciable cross-talk signal. Thus, there is flexibility in picking the fitting function, as long as the samples that are present are appropriately represented. If we Fourier transform the impulse response we obtain a crosstalk MTF component. We then cascade this crosstalk MTF component with other system MTF contributions such as footprint and sampling MTFs.

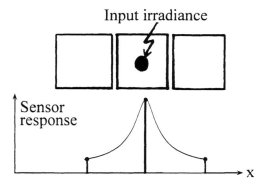

Figure 2.22 Modeling a crosstalk response function. Only the central sensor is illuminated.

For example, charge-transfer inefficiency[15] is seen in charge-coupled devices (CCDs). It is caused by incomplete transfer of charge packets along the CCD delay line. A smearing of the image occurs that is spatial-frequency dependent. The image artifact seen is analogous to a motion blur in the along-transfer direction. For n charge transfers, ε fractional charge left behind at each transfer, and Δx pixel spacing, the crosstalk MTF from charge-transfer inefficiency (Fig. 2.23) is given by

$$\mathrm{MTF}(\xi) = e^{-n\varepsilon[1-\cos(2\pi\xi\Delta x)]} \quad \text{for } 0 \le \xi \le \frac{1}{2\Delta x} \quad . \qquad (2.16)$$

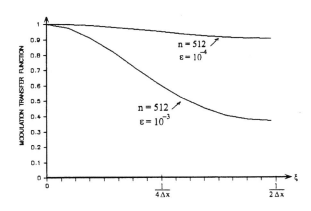

Figure 2.23 Crosstalk MTF caused by charge transfer inefficiency.

Charge-carrier diffusion also leads to crosstalk effects. The absorption of photons in a semiconductor material is wavelength dependent. The absorption is high for short-wavelength photons and it decreases for longer-wavelength photons as the band-gap energy of the material is approached. With less absorption, long-wave photons penetrate deeper into the material, and thus their photogenerated charges must travel further to be collected. The longer propagation path leads to more charge-carrier diffusion, and hence more charge-packet spreading, and ultimately poorer MTF for long-wavelength illumination. A typical family of carrier-diffusion MTF curves for various wavelengths is shown in Fig. 2.24.

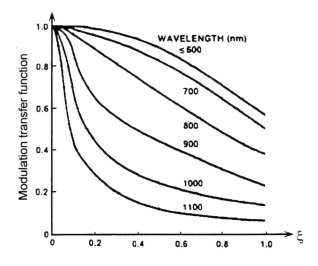

Figure 2.24   Example of variation of carrier-diffusion MTF with illumination wavelength.

2.4 Electronic-network MTF

Electronic networks are essential to electro-optical imagers. They are present in data-acquisition, signal-processing, and display subsystems, and establish a baseline noise level. To cast the electronics transfer function as an MTF and cascade that with the MTFs for other subsystems, we must convert temporal frequency (Hz) into spatial frequency. These frequencies are related by a quantity having units of an effective scan velocity $\mathcal{V}_{scan}$, in either image-plane spatial frequency or object-space angular spatial frequency:

$$f\,[\text{Hz}] = \mathcal{V}_{scan,\text{image-plane}}\,[\text{mm/s}] \times \xi\,[\text{cy/mm}] \qquad (2.17)$$

or

$$f\,[\text{Hz}] = \mathcal{V}_{scan,\text{angular}}\,[\text{mrad/s}] \times \xi\,[\text{cy/mrad}]\,. \qquad (2.18)$$

We can easily visualize the meaning of scan velocity for a scanned-sensor system, such as that seen in Fig. 2.1 because the IFOV is actually moving across the object plane. It is not as easy to visualize scan velocity for a staring system like Fig. 2.2, because there is no motion of the IFOV. However, we can calculate a quantity having units of scan velocity if we know the field of view and the frame rate. However, it is not necessary to explicitly find the scan velocity in order to convert from temporal to spatial frequencies. We only need to find one correspondence, which can be easily done using an electronic spectrum analyzer, as seen in Fig. 2.25. First, we set up a bar target of known fundamental frequency which will make a spatial frequency in the image that we can calculate (knowing the optical magnification), or measure directly from the output signal (knowing the pixel-to-pixel spacing of the detector array). Inputting the output video signal from the detector array to the spectrum analyzer will give a readout of the electrical frequency corresponding to the fundamental image-plane spatial frequency of the bar target.

The shape of the electronics MTF has considerable flexibility. It need not be maximum at dc and it can have sharp cutoffs. Unlike an optics MTF, it is not bounded by an autocorrelation. The addition of a boost filter in the electronics subsystem can partially compensate for the MTF of the optics. A boost filter increases the overall response of the system over a range of frequencies relative to the response at dc. From Chapter 1 we know that classical MTF theory would say that any modulation depth, no matter how small, can be recovered, and hence that a boost filter can compensate for all MTF losses. Practically speaking, the useful application of a boost filter is naturally limited by the effects of electronics noise. Boost filters increase the electrical noise-equivalent bandwidth and hence decrease the image signal-to-noise ratio (SNR).[16] Both high MTF and high SNR are important for good image detectability, so in the design of a proper boost filter we must decide how much boost gain to use, and at what frequencies.

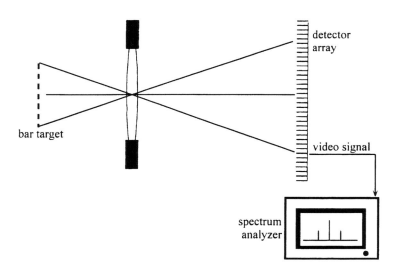

Figure 2.25 Determining temporal- to spatial-frequency conversion.

An image-quality criterion that can be used to quantify this tradeoff is the MTF area (MTFA), which has been field validated to correlate well with image detectability.[17] MTFA is the area between the MTF curve and the noise-equivalent modulation (NEM) curve, which characterizes the system noise in MTF terms. The NEM curve is defined as the amount of modulation depth needed to yield a SNR of unity. Because electronics noise is typically frequency dependent, the NEM is often seen to be a function of spatial frequency. NEM is also referred to as the demand modulation function or the threshold-detectability curve. A convenient representation is to plot the MTF and the NEM on the same graph, as seen in Fig. 2.26. The limiting resolution is the spatial frequency where the curves cross.

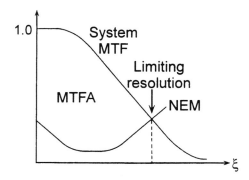

Figure 2.26 Relationship of MTF, NEM, and MTFA.

The power spectral density (PSD) is a common way of describing the frequency content of the electronics noise. The PSD uses units of W/Hz. Because the PSD is in terms of power and the NEM, being in modulation-depth units, is proportional to voltage, we can relate NEM and PSD by

$$\text{NEM}(\xi) = C(\xi)\sqrt{\text{PSD}(\xi)} \quad , \tag{2.19}$$

where the proportionality factor $C(\xi)$ accounts for the display and observer.

Taking MTFA as the quantity to be maximized, we can write the MTFA before application of a boost filter:

$$\text{MTFA}_{\text{before-boost}} = \int_{\xi_1}^{\xi_2}[\text{MTF}(\xi) - \text{NEM}(\xi)]\, d\xi \quad . \tag{2.20}$$

An ideal boost filter amplifies signal and noise equally over its range so, after the boost, the MTFA becomes

$$\text{MTFA}_{\text{after-boost}} = \int_{\xi_1}^{\xi_2}\text{MTF}(\xi)[\text{MTF}(\xi) - \text{NEM}(\xi)]\, d\xi \quad . \tag{2.21}$$

From Eq. (2.21) we see that MTFA is increased, enhancing image detectability, if we require that the range of application of the boost be restricted to those frequencies for which MTF is greater than NEM. The magnitude of the boost that can be practically used is limited by oscillation (ringing) artifacts in the image and by the overall noise level.[18]

2.5 Conclusion

We can apply a transfer-function analysis that was originally developed for classical optical systems to electro-optical systems by generalizing the assumptions of linearity and shift invariance. Linearity is not strictly valid for systems that have an additive noise level, because image waveforms must be of sufficient irradiance to overcome the noise before they can be considered to add linearly. The definition of MTF area (MTFA) allows us to consider a spatial-frequency-dependent signal-to-noise ratio rather than simply a transfer function. Shift invariance is not valid for sampled-data systems but, to preserve the notion of a transfer function, we consider the response of the system to an ensemble average of image waveforms – each with a random position with respect to the array of sampling sites. With the above-mentioned modifications, we can apply a transfer-function approach to a wide range of situations.

References

1. J. Gaskill, *Linear Systems, Fourier Transforms, and Optics*, Wiley, New York (1978).

2. G. D. Boreman and A. Plogstedt, "Spatial filtering by a nonrectangular detector," *Appl. Opt.*, Vol. 28, p. 1165 (1989).

3. K. J. Barnard and G. D. Boreman, "MTF of hexagonal staring focal plane arrays," *Opt. Eng.* 30, p. 1915 (1991).

4. J. Greivenkamp, "Color-dependent optical prefilter for suppression of aliasing artifacts," *Appl. Opt.* 29, p. 676 (1990).

5. E. Dereniak and G. D. Boreman, *Infrared Detectors and Systems*, Wiley, New York (1996), referenced figures reprinted by permission of John Wiley and Sons, Inc.

6. S. K. Park, R. Schowengerdt, and M. Kaczynski, "Modulation-transfer-function analysis for sampled image systems," *Appl. Opt.* 23, p. 2572 (1984).

7. A. Daniels, G. D. Boreman, A. Ducharme, and E. Sapir, "Random transparency targets for modulation transfer function measurement in the visible and IR," *Opt. Eng.* 34, pp. 860-868 (1995).

8. K. J. Barnard, E. A. Watson, and P. F. McManamon, "Nonmechanical microscanning using optical space-fed phased arrays," *Opt. Eng.* 33, p. 3063 (1994).

9. K. J. Barnard and E. A. Watson, "Effects of image noise on submicroscan interpolation," *Opt. Eng.* 34, p. 3165 (1995).

10. L. Huang and U. L. Osterberg, "Measurement of crosstalk in order-packed image-fiber bundles," Proc. SPIE 2536, pp. 480-488 (1995).

11. A. Komiyama and M. Hashimoto, "Crosstalk and mode coupling between cores of image fibers," *Electron. Lett.* 25, pp. 1101-1103 (1989).

12. O. Hadar, D. Dogariu, and G. D. Boreman, "Angular dependence of sampling MTF," *Appl. Opt.* 36, pp. 7210-7216 (1997).

13. O. Hadar and G. D. Boreman, "Oversampling requirements for pixelated-imager systems," *Opt. Eng.* 38, pp. 782-785 (1999).

14. G. D. Boreman and A. Plogstedt, "MTF and number of equivalent elements for SPRITE detectors," *Appl. Opt.* 27, p. 4331 (1988).

15. D. F. Barbe and S. B. Campana, "Imaging arrays using the charge-coupled concept," in *Advances in Image Pickup & Display*, B. Kozan, ed., Academic Press, New York (1977).

16. P. Fredin and G. D. Boreman, "Resolution-equivalent D* for SPRITE detectors," *Appl. Opt.* 34, pp. 7179-7182 (1995).

17. J. Leachtenauer and R. Driggers, *Surveillance and Reconnaissance Imaging Systems*, Artech House, Boston (2001), pp. 191-193.

18. P. Fredin, "Optimum choice of anamorphic ratio and boost filter parameters for a SPRITE based infrared sensor," Proc. SPIE 1488, p. 432 (1991).

Further reading

R. Vollmerhausen and R. Driggers, *Analysis of Sampled Image Systems*, SPIE Press, Bellingham (1998).

CHAPTER 3
OTHER MTF CONTRIBUTIONS

After we consider the fundamental MTF contributions arising from the optics, the sensor array, and the electronics, we must look at the MTF contributions arising from image motion, image vibration, atmospheric turbulence, and aerosol scattering. We briefly consider a first-order analysis of these additional contributions to system MTF; our heuristic approximate approach is a back-of-the-envelope estimate for the image-quality impact of these effects and provides a starting point against which to compare the results of a more advanced analysis.

3.1 Motion MTF

Image-quality degradation arises from movement of the object, image receiver, or optical-system line of sight during an exposure time τ_e. Consider uniform linear motion of the object at velocity v_{obj} and a corresponding linear motion of the image at image velocity v_{img}, as shown in Fig. 3.1. Over an exposure time τ_e, the image has moved a distance $v_{img} \times \tau_e$. This one-dimensional motion blur can be modeled as a rect function

$$h(x) = \text{rect}[x/(v_{img} \times \tau_e)] \, , \qquad (3.1)$$

leading to an MTF along the direction of motion,

$$\text{MTF}_{\text{along-motion}}(\xi) = \left| \frac{\sin(\pi\xi v_{img} \tau_e)}{(\pi\xi v_{img} \tau_e)} \right| \, . \qquad (3.2)$$

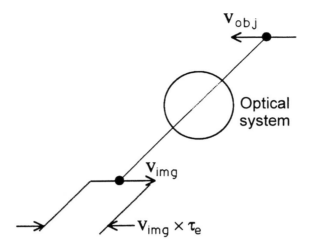

Figure 3.1 Linear motion blur is the product of image velocity and exposure time.

3.2 Vibration MTF

Sometimes the platform on which the optical system is mounted vibrates. The effect of the vibration is typically analyzed one frequency at a time, thus assuming sinusoidal motion. The most important distinction is between high-frequency and low-frequency motion, comparing the temporal period of the motion waveform to the exposure time of the sensors τ_e. The case of high-frequency sinusoidal motion, where many oscillations occur during exposure time τ_e, is the easiest to analyze. As seen in Fig. 3.2, we assume that any object point undergoes sinusoidal motion of amplitude D with respect to the optic axis. The corresponding image-point motion will build up a histogram impulse response in the image plane. This impulse response will have a minimum at the center of the motion and maxima at the image-plane position corresponding to the edges of the object motion, because the object is statistically most likely to be found near the peaks of the sinusoid where the object motion essentially stops and turns around. The process of stopping and turning leads to more residence time of the object at those locations and thus a higher probability of finding the object near the peaks of the sinusoidal motion.

If the sinusoidal object motion has amplitude D, the total width of $h(x)$ is $2D$, assuming unit magnification of the optics. There is zero probability of the image point being found outside of this range, which leads to the impulse response depicted in Fig. 3.3.

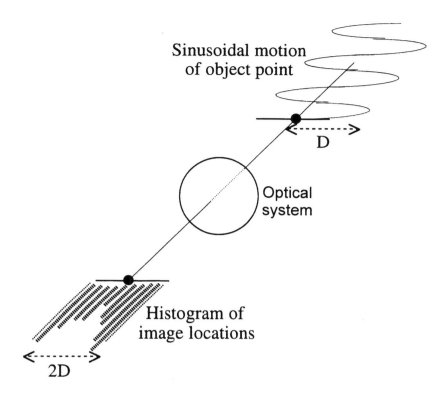

Figure 3.2 High-frequency sinusoidal motion builds up a histogram impulse response in the image plane.

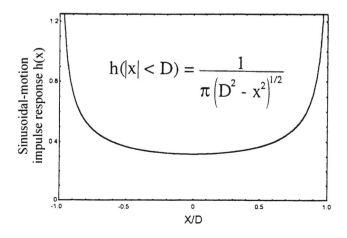

Figure 3.3 Impulse response for sinusoidal motion of amplitude *D*.

If we take the Fourier transform of this $h(x)$, we obtain the corresponding vibration MTF seen in Fig. 3.4.[1]

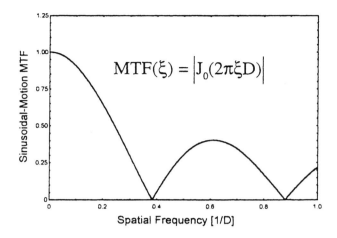

$$MTF(\xi) = \left| J_0(2\pi\xi D) \right|$$

Figure 3.4 MTF for sinusoidal motion of amplitude *D*.

For low-frequency sinusoidal vibrations, the image quality depends on whether the oscillation occurs near the origin or near the extreme points of the object movement during the exposure time. As stated earlier, the velocity slows near the extreme points and is at its maximum near the center of the motion. In the case of low-frequency sinusoidal vibrations, a more detailed analysis is required to predict the number of exposures required to get a single lucky shot where there is no more than a prescribed degree of motion blur.[2]

3.3 Turbulence MTF

Atmospheric turbulence results in image degradation because random refractive-index fluctuations cause ray-direction deviations. We consider a random phase screen (Fig. 3.5), with an autocorrelation width w that is the size of the refractive-index eddy and a phase variance σ^2. The simplest model is one of cloudlike motion where the phase screen moves with time but does not change form: the frozen-turbulence assumption. We look for average image-quality degradation over exposure times that are long compared with the motion of the phase screen.

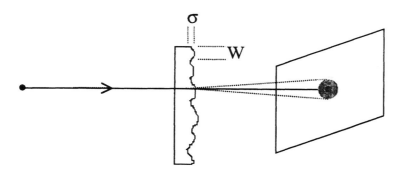

Figure 3.5 Frozen turbulence model for a phase screen.

We consider diffraction from these turbulence cells to be the beam-spreading mechanism and we assume $w \gg \lambda$. Typically the autocorrelation width w is in the range 1 cm $< w <$ 1 m. As seen in Fig. 3.6, the impulse response $h(x)$ consists of a narrow central core from the unscattered radiation and a wide diffuse region from the scattered radiation. Larger phase variation in the screen leads to more of the impulse-response power being contained in the scattered component.

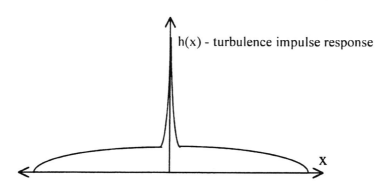

Figure 3.6 Impulse-response form for atmospheric turbulence.

The narrow central core of the impulse response contributes to a broad flat MTF at high frequencies. The broad diffuse scattered component of the impulse response will contribute to an MTF rolloff at low frequencies. We can write[3] the turbulence MTF as

$$\mathrm{MTF}(\xi) = \exp\left(-\sigma^2\left\{1 - \exp\left[-\left(\frac{\lambda\xi}{w}\right)^2\right]\right\}\right), \qquad (3.3)$$

where ξ is the angular spatial frequency in cycles per radian. This MTF is plotted in Fig. 3.7, with the phase variance as a parameter. For a phase variance near zero, turbulence contributes no image-quality degradation because most of the light incident on the phase screen passes through unscattered. As the phase variance increases, more light will be spread into the diffuse halo of the impulse response that we see in Fig. 3.6, and the MTF will decrease at low frequencies. For all of the curves in Fig. 3.7, the MTF is flat at high frequencies. The atmospheric turbulence MTF is only one component of the MTF product and the high-frequency rolloff typically seen for overall system MTF will be caused by some other MTF component.

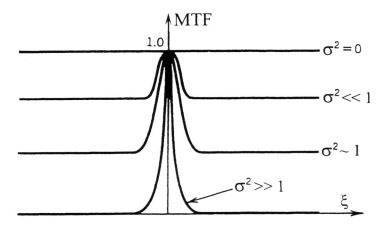

Figure 3.7 Turbulence MTF, parameterized on phase variance (adapted from Ref. 3).

For large phase variance, the turbulence MTF of Eq. (3.3) reduces to a Gaussian form

$$\mathrm{MTF}(\xi) = \exp\left[-\sigma^2\left(\frac{\lambda\xi}{w}\right)^2\right], \tag{3.4}$$

with a 1/e rolloff frequency of $\xi_{1/e} = w/(\lambda\sigma)$. In this limit it is straightforward to identify the issues affecting image quality. The transfer function has a higher rolloff frequency (better image quality) for larger eddy size w (less diffraction), smaller λ (less diffraction), and smaller σ (less phase variation).

3.4 Aerosol-scattering MTF

Forward scattering from airborne particles (aerosols) also causes image degradation. We consider a volume medium with particles of radius a, as seen in Fig. 3.8, and assume that the particle concentration is sufficiently low that multiple-scattering processes are negligible. We consider diffraction from the particles to be the primary beam-spreading mechanism.[4,5] According to Eq. (3.5), attenuation of transmitted beam power ϕ is caused by both absorption (exponential decay coefficient = A) and scattering (exponential decay coefficient = S). Thus, the 1/e distance for absorption is $1/A$, and for scattering is $1/S$:

$$\phi(z) = \phi_0 e^{-(A+S)z} \quad . \tag{3.5}$$

Absorption is not a diffraction process and does not depend on spatial frequency. Because when we set MTF($\xi = 0$) = 1 the absorption effects will be normalized out, only the scattering process is important for development of an aerosol MTF.

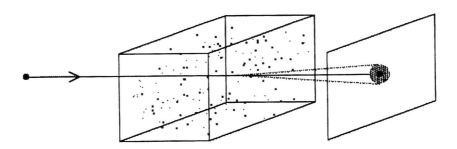

Figure 3.8 Forward-scattering model for aerosols.

We typically assume that the particle size $a > \lambda$. The range of particle sizes of interest for aerosols is approximately 100 μm $< a <$ 1 mm, so the size of the diffracting object a for scattering is smaller by an order of magnitude or more than it was w for the case of atmospheric turbulence. As $a \approx \lambda$, which is important only for very small particles and in the infrared portion of the spectrum, the scattering is nearly isotropic in angle, rather than in the forward direction. This isotropic scattering gives very poor image-transfer quality.

Within the limitation of the forward-scatter assumptions, the impulse response has a narrow central core from the unscattered radiation, along with a diffuse scattered component for turbulence similar to that seen in Fig. 3.6. The only change in the impulse response is that the spatial scale of the halo is typically wider for aerosol scattering than for the case of turbulence, because of the smaller size of the scattering particles. The aerosol-scattering MTF has two functional forms, one corresponding to the low-frequency rolloff region resulting from the Fourier transform of the broad halo, and one corresponding to the flat high-frequency region resulting from the transform of the central narrow portion of the impulse response. The transition spatial frequency ξ_t marks the boundary between the two functional-behavior regions. For the aerosol MTF we have

$$MTF(\xi) = \exp\{-Sz(\xi/\xi_t)^2\}, \ \text{for} \ \xi < \xi_t \tag{3.6}$$

and

$$MTF(\xi) = \exp\{-Sz\}, \ \text{for} \ \xi > \xi_t, \tag{3.7}$$

where z is the propagation distance and the transition frequency is

$$\xi_t = a/\lambda \tag{3.8}$$

in angular [cy/rad] units. Examining Eqs. (3.6) through (3.8) we find that the scattering process is independent of spatial frequency beyond ξ_t. For longer paths (larger z), the MTF decreases at all frequencies, as shown in Fig. 3.9. For more scattering (large S), MTF decreases at all frequencies, as shown in Fig. 3.10.

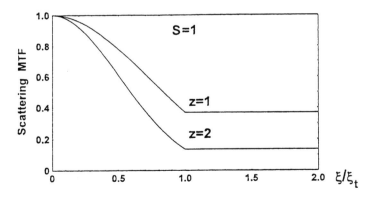

Figure 3.9 Increasing propagation path decreases the aerosol-scattering MTF at all frequencies.

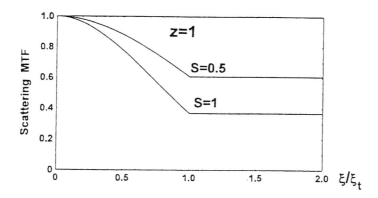

Figure 3.10 Increasing the scattering coefficient decreases the aerosol-scattering MTF at all frequencies.

The transition frequency of Eq. (3.8) is the most important parameter for the scattering MTF because it scales any physical spatial frequency in the ratio in the exponent of Eq. (3.6). Shorter wavelengths and larger particles yield less diffraction and thus result in better image quality; the imaging process behaves in a manner closer to the laws of geometrical optics.

This chapter has provided a brief consideration of MTF contributions from motion, vibration, turbulence, and aerosols. It is intended as a first-order approximate approach to an admittedly complex topic.

References

1. O. Hadar, I. Dror, N. S. Kopeika, "Image resolution limits resulting from mechanical vibrations - Part IV," *Opt. Eng.* 33, p. 566 (1994).

2. D. Wulich and N. S. Kopeika, "Image resolution limits resulting from mechanical vibrations," *Opt. Eng.* 26, p. 529 (1987).

3. J. Goodman, *Statistical Optics*, Wiley, New York (1985), referenced figure reprinted by permission of John Wiley and Sons, Inc.

4. Y. Kuga and A. Ishimaru, "MTF and image transmission through randomly distributed spherical particles," *JOSA A* 2, p. 2330 (1985).

5. D. Sadot and N. S. Kopeika, "Imaging through the atmosphere: practical instrumentation-based theory and verification of aerosol MTF," *JOSA A* 10, p. 172 (1993).

Further reading

N. Kopeika, *A System Engineering Approach to Imaging*, SPIE Press, Bellingham (1998).

CHAPTER 4
MTF MEASUREMENT METHODS

In this chapter, we develop the mathematical relationships between the data and the MTF for the point-spread function (PSF), line-spread function (LSF), and edge-spread function (ESF). One item of notation in this section is that we use * to denote a one-dimensional convolution and ** to denote a two-dimensional convolution. We also consider the relationship between measurement data and MTF for random-noise targets and bar targets.

We begin with a useful approximate formula. For a uniformly distributed blur spot of full width w, the resulting MTF, shown in Fig. 4.1 is

$$\mathrm{MTF}(\xi) = \left| \frac{\sin(\pi\xi w)}{(\pi\xi w)} \right| \quad . \tag{4.1}$$

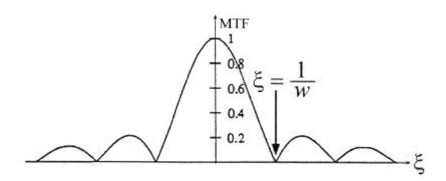

Figure 4.1 MTF for uniform blur spot of width w.

We can use this simple approach as a handy reality check, comparing a measured spot size to calculated MTF values. Often when computers are part of the data-acquisition and data-processing procedures, we cannot check and verify each step of the MTF calculation. In these situations, we can manually verify the computation. First, we visually assess the size of the impulse response to determine the full width of the blur spot by using a measuring microscope – a single-tube combination of objective and eyepiece mounted on a micropositioner spot dimension if we position the crosshairs on one side of the blur spot and note the amount of motion required to place the crosshairs on the other side of the blur spot. Surely the blur spot is not a uniformly distributed irradiance distribution and there is some arbitrariness in the assessment of the full-width criterion in a visual measurement. Nevertheless, we can obtain a back-of-the-envelope estimate for the MTF by taking the visual estimate of w, and then using Eq. (4.1). This MTF is compared to the results of the computer calculation. This comparison will often reveal such errors as missed scale factors of 2π in the spatial-frequency scale, incorrect assumption about wavelength or $F/\#$ in the computer calculation of the diffraction limit, wrong magnification assumed in a reimaging lens, inaccurate assumption about detector-to-detector spacing in the image receiver array, or similar errors. When we compare Fig. 4.1 (using the manual measurement of w) to the calculated MTF, the results should be reasonably close in magnitude (if not in actual functional form), say within about 15%. If this level of agreement is not found, we are well advised to reexamine the assumptions made in the computer calculation before we certify the final measurement results.

4.1 Point-spread function (PSF)

In the idealized arrangement of Fig. 4.2, we use a point source, represented mathematically by a two-dimensional delta function, as the object:

$$f(x,y) = \delta(x,y) \ . \tag{4.2}$$

We assume that the image receiver is continuously sampled; that is, we do not need to consider the finite size of pixels nor the finite distance between samples (we will address these aspects of the instrument response in Chapter 5). Here we assume that we can measure the image-irradiance distribution $g(x,y)$ to the necessary spatial precision. If the object is truly a point source, the two-dimensional image distribution $g(x,y)$ is identically equal to the impulse response $h(x,y)$. This is also called the point-spread function (PSF)

$$g(x,y) = h(x,y) \equiv \text{PSF}(x,y) \ . \tag{4.3}$$

The PSF can be Fourier transformed in two dimensions to yield the two-dimensional OTF. Taking the magnitude yields the MTF

$$\left| \mathcal{FF}[\text{PSF}(x,y)] \right| = \text{MTF}(\xi,\eta) \ . \tag{4.4}$$

This two-dimensional transfer function can be evaluated along any desired profile, for example $\text{MTF}(\xi,0)$ or $\text{MTF}(0,\eta)$.

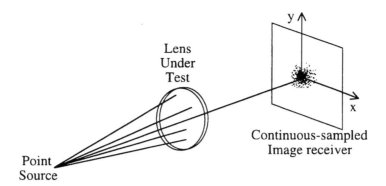

Figure 4.2 Point-spread function measurement configuration.

4.2 Line-spread function (LSF)

Figure 4.3 is a schematic of the measurement setup for the line-spread function (LSF), also called the line response. Instead of the point source used in Fig. 4.2, the LSF test uses a line-source object, which is a delta function in x and a constant in y, at least over a sufficient object height to overfill the field of view of interest in the lens under test:

$$f(x,y) = \delta(x)\ 1(y)\ .\tag{4.5}$$

The two-dimensional image irradiance distribution $g(x,y)$ is the LSF, which is really only a function of one spatial variable (the same variable as that of the impulsive behavior of the line source – in this case, the x direction):

$$g(x,y) \equiv \mathrm{LSF}(x)\ .\tag{4.6}$$

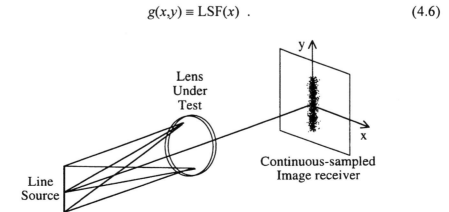

Figure 4.3 Line-spread function measurement configuration.

Each point in the line source produces a PSF in the image plane. These displaced PSFs overlap in the vertical direction and their sum forms the LSF. As seen schematically in Fig. 4.4, the LSF is no more than the two-dimensional convolution of the line source object with the impulse response of the image-forming system:

$$g(x,y) \equiv \text{LSF}(x) = f(x,y) ** h(x,y) = [\delta(x)\,1(y)] ** \text{PSF}(x,y) \;. \qquad (4.7)$$

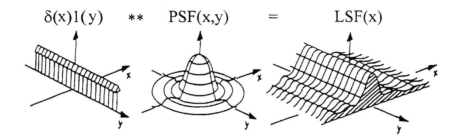

Figure 4.4 The LSF is the two-dimensional convolution of the line source object with the PSF (adapted from Ref. 1).

The y-direction convolution with a constant in Eq. (4.7) is equivalent to an integration over the y direction

$$g(x, y) = \text{LSF}(x) = \int_{-\infty}^{\infty} h(x, y')\,dy' \;, \qquad (4.8)$$

and we verify that the LSF is a function of x alone. It must be independent of y because the object used for the measurement is independent of y. The object, being impulsive in one direction, provides information about only one spatial-frequency component of the transfer function. We can find one profile of the MTF from the magnitude of the one-dimensional Fourier transform of the line response:

$$\left| \mathcal{F}[\text{LSF}(x)] \right| = \text{MTF}(\xi,0) \;. \qquad (4.9)$$

We can obtain other profiles of the transfer function by reorienting the line source. For instance, if we turn the line source by an in-plane angle of 90° we get

$$f(x,y) = 1(x)\,\delta(y) \;, \qquad (4.10)$$

which yields a y-direction LSF that transforms to MTF$(0,\eta)$.

It is important to note that because of the summation along the constant direction in the line-source image, the LSF and PSF have different functional forms. The LSF(x) is not simply the x profile of PSF(x,y):

$$\mathrm{LSF}(x) \neq \mathrm{PSF}(x,0) \ . \tag{4.11}$$

In Fig. 4.5, we compare the PSF and LSF for a diffraction-limited system. We see that while PSF($x,0$) has zeros in the pattern, the LSF(x) does not.

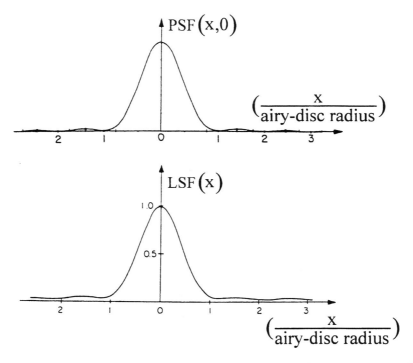

Figure 4.5 Comparison of the *x*-direction functional forms of the PSF and the LSF for a diffraction-limited system. The airy-disc radius is 1.22 λ *F*/# (adapted from Ref. 1).

4.3 Edge-spread function (ESF)

In Fig. 4.6 we see the configuration for the measurement of the edge-spread function. We use an illuminated knife-edge source, a step function, as the object

$$f(x,y) = \mathrm{step}(x)\ 1(y)\ . \tag{4.12}$$

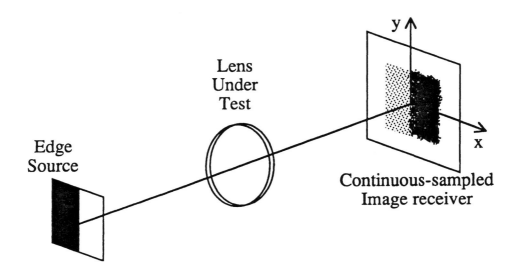

Figure 4.6 Edge-spread-function measurement configuration.

The ESF is the convolution of the PSF with the unit-step function

$$g(x, y) \equiv \text{ESF}(x) = \text{PSF}(x, y) ** \text{step}(x)1(y) \quad . \tag{4.13}$$

The y convolution of the PSF with a constant produces an LSF, and the x convolution with the step function produces a cumulative integration, seen schematically in Fig. 4.7.

$$\text{ESF}(x) = \text{PSF}(x, y) ** \text{step}(x)\, 1(y) = \int_{-\infty}^{x} \text{LSF}(x')dx' \quad . \tag{4.14}$$

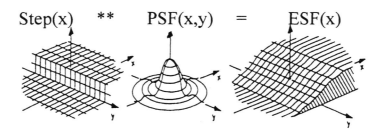

Figure 4.7 The ESF is the two-dimensional convolution of the edge-source object with the PSF (adapted from Ref. 1).

The ESF is a cumulative, monotonically increasing function. Figure 4.8 illustrates the ESF for a diffraction-limited system.

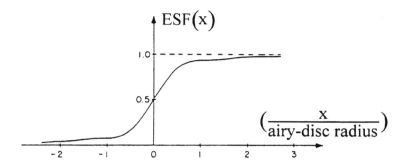

Figure 4.8 Plot of the ESF for a diffraction-limited system. The airy-disc radius is 1.22 λ F/# (adapted from Ref. 1).

We can understand the ESF in terms of a superposition of LSFs.[2,3] Each vertical strip in the open part of the aperture produces a LSF at its corresponding location in the image plane. These displaced LSFs overlap in the horizontal direction and sum to form the ESF. We can write this process as

$$\text{ESF}(x) \approx \sum_{i=1}^{\infty} \text{LSF}(x - x_i) \ . \tag{4.15}$$

In the limit of small displacements, the summation becomes an integral, consistent with Eq. (4.14) To convert ESF data to the MTF, we first take the spatial derivative of the ESF data to invert the integral in Eq. (4.14),

$$\frac{d}{dx}\left\{ \text{ESF}(x) \right\} = \frac{d}{dx} \int_{-\infty}^{x} \text{LSF}(x')\,dx' = \text{LSF}(x), \tag{4.16}$$

seen schematically in Fig. 4.9.

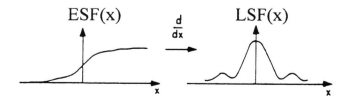

Figure 4.9 Spatial derivative of ESF data produces an LSF (adapted from Ref. 1).

With the LSF in hand, the magnitude of the one-dimensional Fourier transform returns to one profile of the MTF by means of Eq. (4.9). We can obtain any one-dimensional profile of the MTF by re-orienting the knife edge.

4.4 Comparison of PSF, LSF, and ESF

When we compare the advantages and disadvantages of the PSF, LSF, and ESF tests, we find that the PSF test provides the entire two-dimensional OTF in one measurement. The major drawback to the PSF test is that point-source objects often provide too little flux to be conveniently detected. This is particularly true in the infrared portion of the spectrum, where we typically use blackbodies as the flux sources. Flux is not an issue in the visible, because we can use hotter sources with higher flux densities.

We can also use laser sources to illuminate the pinhole. The spatial coherence properties of lasers do not complicate the interpretation of PSF data because a pinhole small enough to produce true PSF data will be spatially coherent across it, whatever the illumination source. However, for LSF and ESF tests, we must ensure that the coherence properties of the laser do not introduce interference artifacts into the data (we will revisit this issue in Chapter 5). The LSF method provides more image-plane flux than does the PSF test. The ESF setup provides even more flux, and has the added advantage that a knife edge avoids slit-width issues. However, the ESF method requires a spatial-derivative operation, which accentuates noise in the data. If we reduce noise by convolution with a spatial kernel, then the data-smoothing operation itself is a contributor to the instrumental MTF.

4.5 Noise targets

Noiselike targets of known spatial-frequency content are useful for MTF testing, particularly for sampled imaging systems such as detector-array image receivers. These noise targets have a random position of the image data with respect to sampling sites in the detector array and measure a shift-invariant MTF that inherently includes the sampling MTF. Noise targets measure the MTF according to

$$\mathrm{PSD}_{\mathrm{img}}(\xi, \eta) = \mathrm{PSD}_{\mathrm{obj}}(\xi, \eta) \times \left[\mathrm{MTF}(\xi, \eta) \right]^2, \qquad (4.17)$$

where PSD denotes power spectral density, defined as the ensemble average of the square of Fourier transform of object or image data. The PSD is a measure of spectral content for random targets. The image PSD is calculated from the image data and, because the object PSD is known beforehand, the MTF can be calculated using Eq. (4.17).

The two main methods for generating noise targets are laser speckle and random transparencies. The laser-speckle method measures detector-array MTF alone because it does not require the use of imaging optics, while random transparencies are used to measure MTF of complete systems, including both the sensor array and the foreoptics.

4.5.1 Laser speckle MTF test

As seen in Fig. 4.10, we generate laser speckle when we pass laser light through an integrating sphere. A dual-slit aperture, with slit separation L, projects random fringes of known spatial frequency onto the detector array.[4]

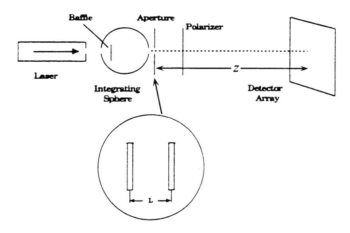

Figure 4.10 Laser-speckle setup for MTF test.

The distance z from the dual-slit aperture to the detector array controls the spatial frequency content of the fringe pattern according to

$$\xi = L/\lambda z , \qquad (4.18)$$

where ξ is the spatial frequency of the narrow high-frequency component of the speckle PSD seen in Fig. 4.11. The baseband component at low spatial frequency is not important for the measurement. This arrangement can measure MTF past the Nyquist frequency of the array because the input of interest is narrowband and, if we know the distance z, we can interpret the pattern even if it is aliased. The solid curve in Fig. 4.11 is the image PSD for a distance z such that the spatial frequency was below the aliasing frequency of the sensor array. The dotted line is the image PSD for a shorter z, for which the spatial frequency of the fringes was above the aliasing frequency. The falloff in MTF between these two frequencies is seen as a decrease in the peak height of the PSD at that value of ξ.

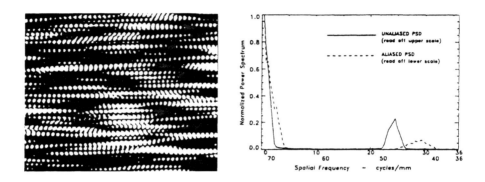

Figure 4.11   Typical narrowband laser-speckle pattern, with the corresponding image PSD. Both aliased and nonaliased PSDs are shown.

4.5.2 Random-transparency MTF test

Figure 4.12 illustrates the setup we use to project a random-transparency object scene into a system under test, which consists of both the imaging optics and a sensor array. We position a transparency at the focus of a collimator and backlight it using a uniform source. An alternate configuration allows for a reflective random object to be imaged.

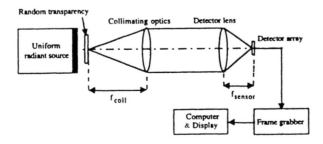

Figure 4.12 Setup for the random-transparency MTF test.

Visible-wavelength transparencies are typically fabricated on photographic film. At infrared wavelengths, materials constraints generally favor electron-beam lithographic processes with chrome on a substrate such as ZnSe. The input power spectrum, which is usually flat, produces a bandlimited white-noise object (Fig. 4.13), although other PSD functions, such as narrowband noise, have been demonstrated.[5]

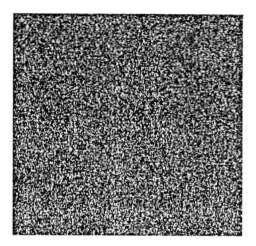

Figure 4.13 Random object with a bandlimited white-noise PSD.

4.6 Bar-target MTF measurement

Accurate sinusoidal targets are difficult to manufacture because of the need for strict harmonic-distortion control and because of materials issues that, at least in the infrared, typically require half-tone electron-beam lithography. Thus, a very common target for measuring MTF is the three-bar or four-bar binary pattern, as seen in Fig. 4.14, which is much easier to fabricate.

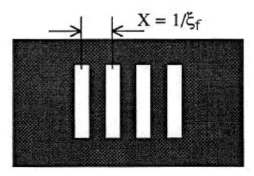

Figure 4.14 Binary four-bar target.

Any particular bar target is specified in terms of its fundamental frequency ξ_f, which is the inverse of the center-to-center spacing of the bars. Typically the widths of the lines and the spaces are equal. Using this type of target, we measure the modulation depth of the image waveform as a function of ξ_f. This measured image modulation depth (IMD) is not exactly the MTF at ξ_f, because of extra frequency components that contribute to the IMD at both higher and lower frequencies than ξ_f. In practice, we remove these other frequency components by computation or filter them out electronically or digitally.

For an infinite square wave, we define the contrast transfer function (CTF) as the image modulation depth as a function of spatial frequency. We can derive a series conversion between CTF and MTF using the Fourier decomposition of the square waves:[6]

$$CTF(\xi_f) = \frac{4}{\pi}\left\{ MTF(\xi = \xi_f) - \frac{MTF(\xi = 3\xi_f)}{3} + \frac{MTF(\xi = 5\xi_f)}{5} - \cdots \right\} \quad (4.19)$$

and

$$MTF(\xi) = \frac{\pi}{4}\left\{ CTF(\xi_f = \xi) + \frac{CTF(\xi_f = 3\xi)}{3} - \frac{CTF(\xi_f = 5\xi)}{5} + \cdots \right\} . \quad (4.20)$$

We can see from Eq. (4.20) that the calculation of MTF at any particular frequency requires that a CTF measurement be made at a series of frequencies that are harmonically related to the desired frequency of the MTF measurement. We must repeat this operation for each frequency at which we want to find the MTF. The number of CTF measurements actually required in practice is limited because the CTF will be negligibly small at the higher harmonic frequencies. Typically we measure the CTF at enough frequencies to plot a smooth curve, and then we interpolate to find the CTFs at the frequencies we need to compute an MTF curve from the CTF data. It is generally not sufficiently accurate to just take the CTF measurements as MTF measurements. A built-in bias makes the CTF higher at all frequencies than the corresponding MTFs; this bias can be readily seen in the first term in the summation of Eq. (4.19), which says that

$$CTF(\xi_f) \approx (4/\pi)\, MTF(\xi_f) , \quad\quad\quad\quad (4.21)$$

so that the CTF at any frequency is generally greater than the MTF, depending on the magnitude of the higher harmonics.

From a measurements point of view, however, we seldom use infinite square waves; the binary bar targets of Fig. 4.14 are preferable. Note that, because their Fourier transform is a continuous function of frequency rather than a discrete-harmonic Fourier series, the series of Eqs. (4.19) and (4.20) are not strictly valid for three- and four-bar targets. To answer the question of the validity of the series conversion, let us compare[7] four different curves (the MTF, CTF, three-bar IMD, and four-bar IMD) for two optical systems of interest: a diffraction-limited circular-aperture system and a diffraction-limited annular-aperture system with a 50% diameter obscuration. If the IMD curves are close to the CTF curve, then we can use the series to convert bar-target data to MTF. If not, we can use filtering to convert bar-target modulation data to MTF.

In Fig. 4.15, we consider the diffraction-limited circular-aperture system. The bias toward higher values of the CTF as compared to the MTF, mentioned previously in connection with Eq. (4.21), is evident. The IMD curves for this case are very close to the CTF curve, and we can use the series conversions of Eqs. (4.19) and (4.20). The few-percent difference would not be measurable in practice.

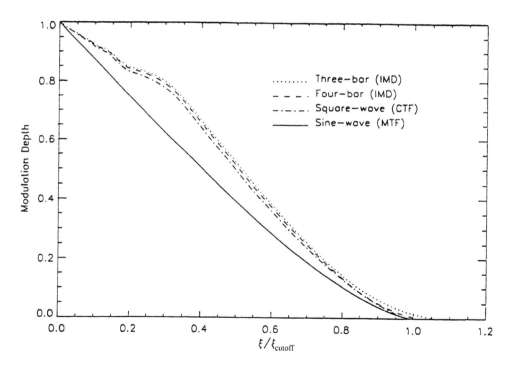

Figure 4.15 Comparison of MTF, CTF, three-bar IMD, and four-bar IMD for a diffraction-limited circular-aperture system.

The situation is a bit different in Fig. 4.16, where we consider the case of an obscured-aperture system. The MTF curve is not as smooth as for the case of no obscuration, and these changes in derivative make the weighting of the different terms of the series a more complicated function of frequency. The CTF curve still exceeds the MTF curve. At some frequencies, the three-bar and four-bar IMD curves are nearly identical to the CTF, and at other frequencies there is perhaps a 10% difference in absolute MTF. For MTF curves with even more abrupt slope changes,[7] the difference between CTF and the IMD curves is larger.

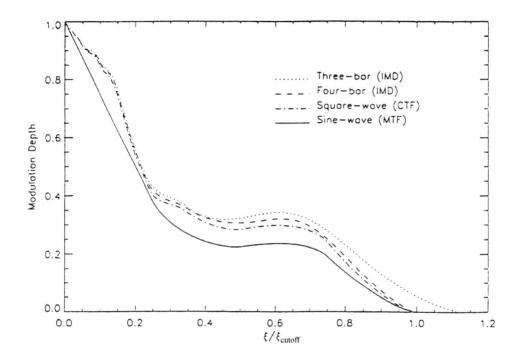

Figure 4.16 Comparison of MTF, CTF, three-bar IMD, and four-bar IMD for a diffraction-limited annular-aperture system with a 50% diameter obscuration.

The previous comparisons do not answer the general question of the accuracy of the series conversion between bar-target data and MTF, except to say that in the case of a diffraction-limited system with no obscuration, the series should be accurate enough. In general, we do not know the form of the MTF curve beforehand, so we developed a data-processing procedure that does not require any assumptions about the smoothness of the MTF curve. When digitized image data are available, a direct bar-target-to-MTF conversion can be performed using an FFT calculation.

Figure 4.17 shows a measured three-bar-target magnitude spectrum $S_{output}(\xi)$, and the corresponding calculated input spectrum $S_{input-bar-target}(\xi)$, both having been normalized to unity at $\xi = 0$. The measured output spectrum is filtered by system MTF, and because MTF decreases with frequency, the peak of output spectrum occurs at lower frequency than fundamental ξ_f of the input target. Without knowing the MTF, we have difficulty determining how much the peak was shifted, so instead we use the zeros of the spectrum (the locations of which are not shifted by the MTF) to determine the fundamental frequency of the input target. We determine ξ_f from the first-zero location. For a three-bar target

$$\xi_f = 3 \times \xi_{\text{first-zero}} \ . \tag{4.22}$$

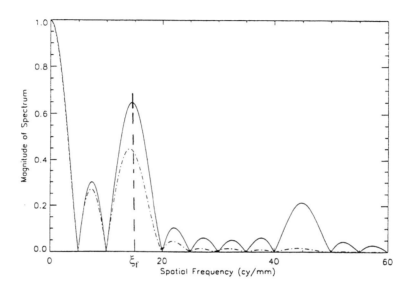

Figure 4.17   Measured three-bar-target magnitude spectrum $S_{\text{output}}(\xi)$ (dotted curve), and the corresponding calculated input spectrum $S_{\text{input-bar-target}}(\xi)$ (solid curve).

If we know ξ_f and know whether the target was a three-bar or a four-bar target, then we can completely determine the normalized magnitude spectrum of the input and we can find the MTF at the fundamental frequency ξ_f of any particular target by

$$\text{MTF}(\xi = \xi_f) = \left[\frac{S_{\text{output}}(\xi = \xi_f)}{S_{\text{input-bar-target}}(\xi = \xi_f)} \right] . \tag{4.23}$$

Equation (4.23) allows us to measure the MTF at the fundamental frequency of the particular target from digitized bar-target data without the need for a series conversion. Even though there is more information present than just at the fundamental frequency, we have found that the accuracy suffers when the frequency range of the measurement is extended beyond the fundamental. Thus, the measurement is still made one frequency at a time, but now without concern about the accuracy of a series conversion, as a digital-filtering technique is used to isolate ξ_f from the continuous spectrum of the bar target.

    This chapter has presented the idealized mathematical relationships among the various data sets and the MTF. In the next chapter we will consider some of the practical issues affecting these measurements.

References

1. J. Gaskill, *Linear Systems, Fourier Transforms, and Optics,* Wiley, New York (1978), referenced figures reprinted by permission of John Wiley and Sons, Inc.

2. B. Tatian, "Method for obtaining the transfer function from the edge response function," *JOSA* 55, p. 1014 (1965).

3. R. Barakat, "Determination of the optical transfer function directly from the edge spread function," *JOSA* 55, p. 1217 (1965).

4. M. Sensiper, G. D. Boreman, A. Ducharme, and D. Snyder, "Modulation transfer function testing of detector arrays using narrow-band laser speckle," *Opt. Eng.* 32, pp. 395-400 (1993).

5. A. Daniels, G. D. Boreman, A. Ducharme, and E. Sapir, "Random transparency targets for modulation transfer function measurement in the visible and IR," *Opt. Eng.* 34, pp. 860-868 (1995).

6. J. Coltman, "The specification of imaging properties by response to a sine wave input," *JOSA* 44, p. 468 (1954).

7. G. D. Boreman and S. Yang, "Modulation transfer function measurement using three- and four-bar targets," *Appl. Opt.* 34, pp. 8050-8052 (1995).

CHAPTER 5
PRACTICAL MEASUREMENT ISSUES

In this chapter, we will consider a variety of practical issues related to MTF measurement, including the cascade property of MTF multiplication, the quality of the auxiliary optics such as collimators and relay lenses, the issue of source coherence, and the finite size of the source for a PSF or LSF test. We will also look at the instrumental-MTF contribution of the detector or the sensor array that receives the image-irradiance data. We will investigate ways to increase the signal-to-noise ratio, making use of redundancy in the source and image planes. We will conclude with some comments about repeatability and accuracy of MTF measurements and the use of computers for data acquisition and processing. At the end of the chapter we will consider four different instrument approaches that are representative of commercial MTF equipment, and identify the design tradeoffs that were made.

5.1 Cascade properties of MTF

We often account for the combination of several subsystems by simply multiplying MTFs. That is, the overall system MTF is calculated as a point-by-point product of the individual subsystem MTFs, as in Eq. (1.9) and Fig. 1.7. This is a very convenient calculation, but it is sometimes not correct. We want to investigate the conditions under which we can multiply MTFs. In consideration of incoherent imaging systems, the MTF multiplication rule can be simply stated: we can multiply MTFs if each subsystem operates independently on an incoherent irradiance image. We will illustrate this rule by examples.

Consider the example seen in Fig. 5.1, the combination of a lens system (curve a) and detector (curve b) in a camera application. The MTF of the combination (curve c) is a point-by-point multiplication of individual MTFs, because surely the detector responds to the spatial distribution of irradiance (W/cm^2) in the image plane without any partial-coherence effects.

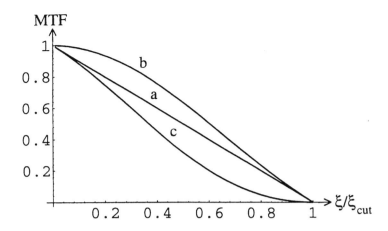

Figure 5.1   Cascade of MTFs of a lens system (curve a) and a detector (curve b) produces the product MTF (curve c).

A relay lens pair with a diffusing screen at the intermediate image plane is another case where we can cascade MTFs by means of a point-by-point multiplication (Fig. 5.2). The exitance (W/cm^2) on the output side of the diffuser is proportional to the irradiance (W/cm^2) at the input face. Any point-to-point phase correlation in the intermediate image is lost in this process. The diffuser forces the two systems to interact independently, regardless of their individual state of correction. The two relay stages cannot compensate for the aberrations of each other because of the phase-randomizing properties of the diffuser. The MTFs of each stage simply multiply one another, and the product MTF is lower than either of the components. This relay-lens example is a bit contrived because we typically do not have a diffuser at an intermediate image plane (from a radiometric point of view, as well as for image-quality reasons), but it illustrates the MTF multiplication rule by presenting the second stage with an incoherent irradiance image formed by the first stage.

A case in which cascading MTFs will not work is illustrated in Fig. 5.3, which is a lens combination for which the second lens balances the spherical aberration of the first lens. Neither lens is well corrected by itself, as seen by the poor individual MTFs. The system MTF is higher than either of the individual-component MTFs, which cannot happen if the MTFs simply multiply. The intermediate image is partially coherent, so these lenses do not interact on an independent irradiance basis.[1] Lens 1 does not simply present an irradiance image to lens 2. Specification of an incoherent MTF for each separate element is not an accurate way to analyze the system because separate specifications do not represent the way that the lenses interact with each other.

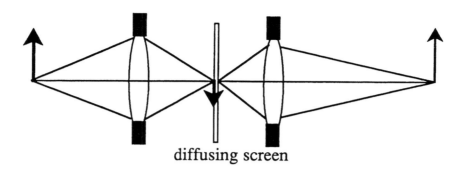

diffusing screen

Figure 5.2   Relay lens pair with diffuser at intermediate image plane. MTF of each stage is simply multiplied.

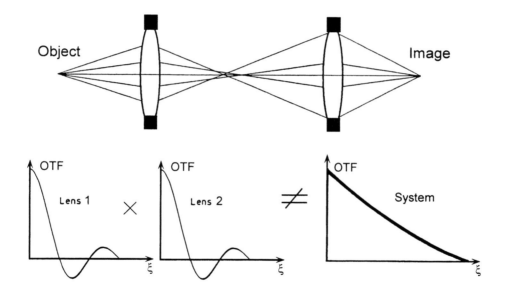

Figure 5.3   Pair of relay lenses for which the MTFs do not cascade.

However, if two systems have been independently designed and independently corrected for aberrations, then the cascade of geometrical MTFs is a good approximation. We take the example seen in Fig. 5.4 of an afocal telescope combined with a final imaging lens, a typical front end in an infrared imager. Each subsystem can perform in a well-corrected manner by itself. As noted in Chapter 1, the limiting aperture (aperture stop) of the entire optics train determines the diffraction MTF. This component does not cascade on an element-by-element basis. Diffraction MTF is included only once in the system

MTF calculation. This distinction is important when measured MTFs are used. For example, suppose that separate measured MTF data are available for both subsystems. Simply multiplying the measured MTFs would count diffraction twice because diffraction is a contributor to any measurement of an individual system.

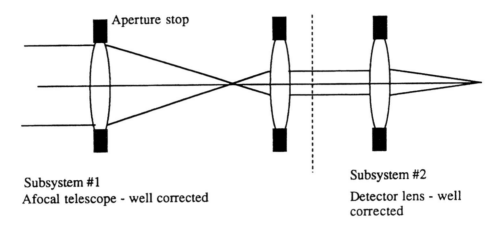

Figure 5.4 Combination of afocal telescope and detector lens.

Because the front end of the afocal in Fig. 5.4 is the limiting aperture of the system, we can use the measured MTF data for subsystem #1:

$$\mathrm{MTF}_{1,\mathrm{meas}}(\xi) = \mathrm{MTF}_{1,\mathrm{geom}}(\xi) \times \mathrm{MTF}_{1,\mathrm{diff}}(\xi) \ . \qquad (5.1)$$

To find the geometrical MTF of subsystem #2, we must separate the geometrical and diffraction contributions[2] in the measured MTF of the subsystem:

$$\mathrm{MTF}_{2,\mathrm{meas}}(\xi) = \mathrm{MTF}_{2,\mathrm{geom}}(\xi) \times \mathrm{MTF}_{2,\mathrm{diff}}(\xi) \qquad (5.2)$$

and

$$\mathrm{MTF}_{2,\mathrm{geom}}(\xi) = \frac{\mathrm{MTF}_{2,\mathrm{meas}}(\xi)}{\mathrm{MTF}_{2,\mathrm{diff}}(\xi)} \ . \qquad (5.3)$$

The only diffraction MTF that contributes to the calculation of the total MTF is that of subsystem #1, so for the total system MTF we have

$$\mathrm{MTF}_{\mathrm{total}}(\xi) = \mathrm{MTF}_{1,\mathrm{diff}}(\xi) \times \mathrm{MTF}_{1,\mathrm{geom}}(\xi) \times \mathrm{MTF}_{2,\mathrm{geom}}(\xi) \ . \qquad (5.4)$$

5.2 Quality of auxiliary optics

Figure 5.5 is a schematic of the typical MTF setup where the unit under test (UUT) images a point-source object.

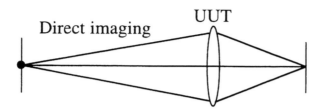

Figure 5.5   MTF test where source is directly imaged by the unit under test.

However, in practice, auxiliary optical elements are often used in the setup, as seen in Fig. 5.6. For instance, additional optics are used to simulate a target-at-infinity condition, to magnify a small image-irradiance distribution before acquiring it with a detector array, or to test an afocal element.

To prevent the auxiliary optics from impacting the MTF results, the following conditions must be met. First, the aperture of the collimator must overfill that of the UUT so that all aperture-dependent aberrations of the UUT are included in the measurement. Second, in the case of the re-imager, the relay-lens aperture must be sufficiently large that it does not limit the ray bundle focused to the final image plane. In both instances, the aperture stop of the end-to-end measurement system must be at the unit under test.

We want the auxiliary optics to be diffraction limited; that is, the geometrical-aberration blur of the auxiliary optics should be small compared to its diffraction blur. Because the previous condition required that the auxiliary optics overfill the aperture of the UUT – putting the aperture stop at the UUT – the diffraction limit will occur in the UUT also. If the auxiliary-optics angular aberrations are small compared to the diffraction blur angle of the auxiliary optics, then the aberrations of the auxiliary optics will be even smaller compared to the larger diffraction blur angle of the UUT. Under these conditions, the image blur of the UUT can be measured directly, because both the diffraction and aberration blur angles of the auxiliary optics are small compared to the UUT blur. If the auxiliary optics is not diffraction-limited, then its aberration blur must be characterized and accounted for in the MTF product that describes the measured data. Then, this instrumental aberration contribution can be divided out by the method considered in Section 5.4, providing that the aberrations of the auxiliary optics are small enough not to seriously limit the spatial-frequency range of the measurement.

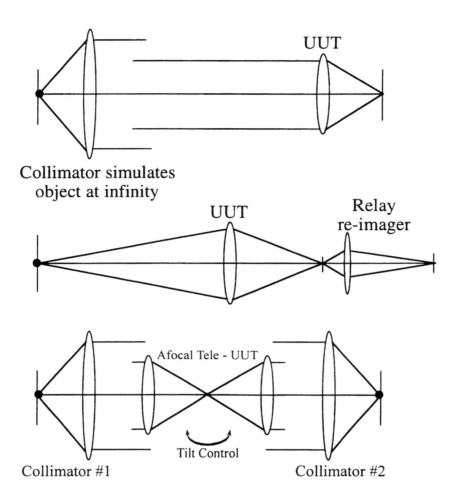

Figure 5.6 Auxiliary optics used in MTF-test setups.

5.3 Source coherence

In MTF tests that involve the imaging of a back-illuminated extended target (for instance, the LSF, ESF, or bar-target tests), the illumination must be spatially incoherent to avoid interference artifacts between separate locations in the image that can corrupt MTF data. If the extended source is itself incandescent (as in a hot-wire LSF test for the infrared), the source is already incoherent. When an extended aperture is back illuminated, coherence effects become important. When incandescent-bulb sources are used with a condenser-lens setup to back illuminate the source aperture, partial-coherence interference effects can be present in the image. The usual way reduce the source coherence for an incandescent source is to place a ground-glass diffuser on the source side, adjacent to the aperture being illuminated, as seen in Fig. 5.7. The phase differences between various correlation cells of the diffuser are generally sufficient for reduction of the degree of partial coherence of the source.

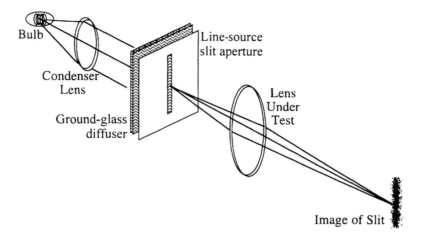

Bulb

Line-source
slit aperture

Condenser
Lens

Lens
Under
Test

Ground-glass
diffuser

Image of Slit

Figure 5.7 A ground-glass diffuser is used to reduce the partial coherence of an incandescent source.

If we illuminate the slit aperture of Fig. 5.7 with a collimated laser beam, the spatial coherence among the points on the object is nearly unity, and we will need a more elaborate setup to reduce the coherence. Again, we will use a ground-glass diffuser at the illuminated aperture, but we must move the diffuser to affect a speckle-averaging function. Mechanical actuators are used to move the diffuser on a time scale that is short compared to the integration time of the sensor that acquires the final image data.

Concerns about the spatial coherence of the input object generally do not apply to pinholes, because if a source is small enough to behave as a point source with respect to the system under test, it is essentially only a single-point emitter, which is self-coherent by definition. We consider spatial coherence only when adjacent point sources might interfere with each other. A pinhole aperture can be illuminated by a laser source without appreciable spatial-coherence artifacts.

5.4 Correcting for finite source size

Source size is an issue for PSF and LSF measurements. We must have a finite source size to have a measurable amount of flux. For a one-dimensional analysis, recall Eq. (1.1), where f, g, and h are the object, image, and impulse-response irradiance distributions, respectively:

$$g(x) = f(x) * h(x) , \qquad (5.5)$$

and Eq. (1.5), where F, G, and H are the object spectrum, the image spectrum, and the transfer function, respectively:

$$G(\xi) = F(\xi) H(\xi) . \qquad (5.6)$$

If input object $f(x) = \delta(x)$, then the image $g(x)$ is directly the PSF $h(x)$:

$$g(x) = \delta(x) * h(x) = h(x) . \qquad (5.7)$$

For a delta-function object, the object spectrum $F(\xi) = 1$, a constant in the frequency domain, and the image spectrum $G(\xi)$ is directly the transfer function $H(\xi)$:

$$G(\xi) = F(\xi)H(\xi) = H(\xi) . \qquad (5.8)$$

Use of a non-delta-function source $f(x)$ effectively bandlimits the input source spectrum. A narrow source $f(x)$ implies a wide source spectrum $F(\xi)$, and a wider source implies a narrower spectrum. The source spectrum $F(\xi)$ will fall off at high frequencies rather than remain constant. Usually, a one-dimensional rect function is convenient to describe the source width and the corresponding sinc function for the source spectrum:

$$f(x) = \text{rect}(x/w) \qquad (5.9)$$

and

$$F(\xi) = \sin(\pi\xi w)/(\pi\xi w) . \qquad (5.10)$$

In the case of a non-delta-function source, we need to divide the measured image spectrum by the object spectrum to solve Eq. (5.6) for $H(\xi)$:

$$H(\xi) = [G(\xi)] / [F(\xi)] , \qquad (5.11)$$

which is equivalent to a deconvolution of the source function from the image data. Obtaining the transfer function by the division of Eq. (5.11) would be straightforward if not for the effects of noise. We cannot measure the image spectrum $G(\xi)$ directly; the measured output spectrum is the image spectrum added to the noise spectrum

$$G_{\text{meas}}(\xi) = G(\xi) + N(\xi) , \qquad (5.12)$$

where the noise spectrum $N(\xi)$ is defined as the square root of the power spectral density (PSD) of the electronics noise

$$N(\xi) = \sqrt{\text{PSD}(\xi)} \quad . \tag{5.13}$$

The division of Eq. (5.11) becomes

$$H(\xi) = \frac{G_{\text{meas}}(\xi)}{F(\xi)} = \frac{G(\xi) + N(\xi)}{F(\xi)} = H(\xi) + \frac{N(\xi)}{F(\xi)} \ , \tag{5.14}$$

which yields valid results at spatial frequencies for which

$$F(\xi) \gg N(\xi) \ , \tag{5.15}$$

so that the last term in Eq. (5.14) is negligible. For frequencies where the input spectrum is near zero, the deconvolution operation divides the finite noise spectrum by a very small number, and the calculated MTF will blow up as seen in Fig. (5.8).

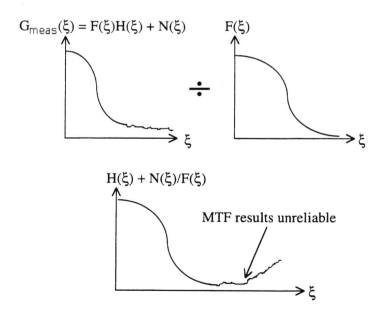

Figure 5.8 Once the source spectrum has fallen to the level of the noise spectrum, MTF test results are invalid because of division-by-zero artifacts.

The MTF data are obviously not valid for these frequencies. To extend the frequency range of the test as far as possible, we want a wide source spectrum, which requires a narrow input source. There is a practical tradeoff here, because a smaller source gives less flux and poorer signal-to-noise ratio. We want to use a source that is sufficiently narrow that the source spectrum is appreciable at the upper end of the spatial-frequency band of interest. If a small source yields a poor signal-to-noise ratio, then brighter illumination is required to increase the source spectrum at all frequencies, compared to the noise level of the electronics (which we assume to be independent of illumination level).

5.5 Correcting for the image-receiver MTF

In any MTF test, we need an image receiver to acquire the image-irradiance function $g(x,y)$. We would like to perform this image acquisition on a continuous basis, with infinitely small pixels and infinitely small spatial-sampling intervals, which were the conditions for the development of the PSF, LSF, and ESF in Chapter 4 (specifically in Figs. 4.2, 4.3, and 4.6). However, from a practical viewpoint, there is always a finite-sized averaging area for an image-plane sensor and a finite distance between data samples. These finite dimensions of the image receiver also contribute to the MTF of the test instrument, and those contributions must be divided out from the measured data (in a similar manner to Eq. (5.11)) in order for us to obtain the MTF of the unit under test.

5.5.1 Finite pixel width

The detector used to acquire the image can be staring or scanning, but the finite dimension of its photosensitive area results in a convolution of the image irradiance $g(x)$ with the pixel footprint impulse response $h_{\text{footprint}}(x)$. The MTF component of finite-footprint pixels is the same as discussed previously in Section 2.1. For most measurement situations, we use a one-dimensional rect function of width w to describe the pixel footprint, leading to the MTF seen in Eq. (2.2):

$$\text{MTF}_{\text{footprint}}(\xi) = \left| \text{sinc}(\xi w) \right| = \left| \frac{\sin(\pi\xi w)}{\pi\xi w} \right| . \qquad (5.16)$$

This term is a component of the instrumental MTF that should be divided out of the measured MTF to obtain the MTF of the unit under test.

5.5.2 Finite sampling interval

For a staring-array image receiver, the sampling interval is the center-to-center pixel spacing of the array. For a scanning sensor, the sampling interval is the spatial distance at which successive samples are taken. We can determine this sampling distance if we know the sampling-time interval used to sample the analog video waveform, using the image-plane scan velocity to convert from time to distance units, in a manner similar to Eqs. (2.17) and (2.18). For manually scanned sensors such as a photomultiplier tube on a micropositioner, the sampling interval is simply how far we move the sensor between samples.

We can apply Eq. (2.12) to determine the sampling MTF, which again represents an instrumental-MTF component that should be divided out of the final MTF results. However, note that sampling-MTF effects are not always present in the data, and thus we do not always need to divide them out. The sampling MTF was derived on the assumption of a random position of the image with respect to the sampling sites. Test methods that use the spatial-noise targets discussed in Chapter 4.5 acquire MTF-test data on an ensemble of images with no particular alignment. These tests include the sampling MTF as part of the measurement. In these situations, the sampling MTF is part of the instrumental MTF and must be divided out when we calculate the MTF of the unit under test. In contrast, the measurement procedures for PSF, LSF, ESF, and bar-target tests typically involve alignment of the image with respect to sampling sites. We tweak the alignment to produce the best-looking image, typically observing the data on a TV monitor or an oscilloscope as we perform the alignment. In this case, sampling MTF does not filter the image data, and any removal of a sampling MTF will unrealistically bias the MTF measurement toward higher values. Even with this fine alignment of the target with respect to the sampling sites, the deterministic targets (PSF, LSF, ESF, bar target) are typically only accurate to $\xi \approx 1/(4\Delta x_{samp})$ – half the Nyquist frequency – because of aliasing artifacts. For example, if we measure the modulation depth of a bar target, struggle with the alignment, and cannot get all four bars to have the same height on the display, then we are probably trying to measure MTF at too high a frequency, and the measurement of the modulation depth will not be accurate. Remember that the bar targets contain a range of frequencies, not just the fundamental, and even if the fundamental is not aliased, some of the higher frequencies will be. This bar-target aliasing effect was seen in Fig. 2.8. Targets used for PSF, LSF, and ESF contain very high frequencies also; thus aliasing artifacts will limit the accuracy of these tests as well for frequencies above half-Nyquist.

If we want to measure MTF at higher frequencies, then we might opt to use an oversampled superresolution knife-edge test.[3] This test requires us to introduce a known slight tilt to the knife edge. The data sets from each line are interleaved with an appropriate subpixel shift, corresponding to the actual position of the knife edge with respect to the column, to create a LSF measurement with a very fine sampling interval. The other option is to use a narrowband noise-target test, such as the laser speckle MTF test (Fig. 4.10) or the random-transparency test with a similar narrowband PSD. These can be used to measure MTF out to twice the Nyquist frequency because the narrowband nature of the noise permits us to interpret the data even in the presence of aliasing.

5.6 Summary of instrumental MTF issues

In a typical test scenario, we have a number of impulse responses that convolve together to give the measured PSF: the diffraction of the unit under test, the aberration of the unit under test, the aberration of the collimator, the detector footprint, the source size, and the sampling. We want to isolate the PSF of the unit under test, which is

$$PSF_{UUT} = PSF_{UUT,diff} * PSF_{UUT,aberr} \ . \tag{5.17}$$

The other PSF terms should be known and much narrower than the PSF of the UUT, so that we can divide them out in the frequency domain without limiting the frequency range of the MTF test.

5.7 Increasing signal-to-noise ratio in PSF, LSF, and ESF tests

We can use a variety of image-averaging techniques to increase the signal-to-noise ratio (SNR) in PSF, LSF, and ESF tests. These averaging techniques can be implemented in either in the object plane or in the image plane, provided we assume that the data is the same in one direction and average over that direction. The SNR usually increases as square root of the number of independent samples, so we stand to gain considerable SNR advantage if we can effectively average a large number of samples (such as the number of rows of a typical CCD array). In certain instances (such as operation in the amplifier-noise limit), the SNR can grow as fast as linearly with respect to the number of samples.

5.7.1 Object- and image-plane equivalence

This technique requires that we assume that we match the object and receiver symmetry for optimum flux collection. In this section, we equate higher flux collection with better SNR. Note, however, that larger detectors generally exhibit more noise than smaller detectors; the root-mean-square (rms) noise is proportional to the square root of the sensor area. This dependence on area reduces the SNR gain somewhat. But, because the collected power (and hence the signal) is proportional to the detector area and the rms noise is proportional to the square root of the detector area, the more flux we can collect, the better our measurement SNR, even if that means we must use a larger detector.

In the configurations illustrated in Figs. 4.2, 4.3, and 4.6, we postulated a continuously sampled image receiver analogous to a point receiver that scans continuously, equivalent to a densely spaced array of point receivers. If we want to obtain PSF data, our test setup must include a point source and a point receiver. The only option to increase SNR if we use a PSF test setup for us to increase the source brightness or to average over many data sets.

If, however, an LSF measurement is sufficient, we can accomplish the measurement in a number of ways – some of which yield a better SNR than others. We can use a linear source and a point receiver, such as the configuration seen in Fig. 4.3. This system will give us a better SNR than PSF measurement because we are utilizing a larger-area source. Equivalently, as far as the LSF data set is concerned, we can use the configuration seen in Fig. 5.9, a point source and a slit detector (or a slit in front of a large-area detector). The data acquired are equivalent to the data for a LSF test, because of the vertical-direction averaging. Just like the situation using a line source and a point receiver, this collects more flux than a PSF test and has a better SNR. However, as long as we are using a linear detector, we might also want to use a linear source (Fig. 5.10). The data still provide an LSF test, but now the source and the receiver have the same geometry. This configuration collects the most flux and will provide the best SNR of any LSF test setup.

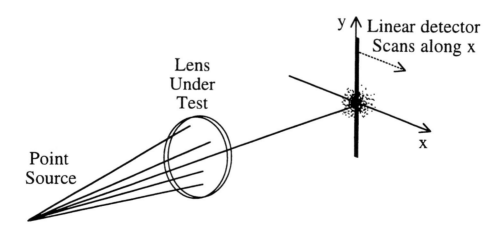

Figure 5.9   A PSF test performed with a linear receiver produces data equivalent to an LSF test.

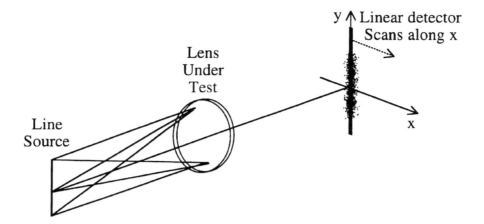

Figure 5.10 An LSF test performed with a linear receiver produces a better SNR than using a point receiver.

A number of different configurations will work for ESF tests, and some are better than others from a SNR viewpoint. We begin with a knife-edge source and a scanned point receiver (Fig. 4.6). This system configuration is equivalent to the configuration of Fig. 5.11 from a flux-collection viewpoint, where the ESF measurement is performed with a point source and a knife edge in front of a large detector as the image receiver. We will obtain a better SNR (and the same ESF data), using the setup illustrated in Fig. 5.12, which utilizes a line source and a scanning knife edge in front of a large detector. We could also use a knife-edge source and a scanning linear receiver, because the data set is constant in the vertical direction.

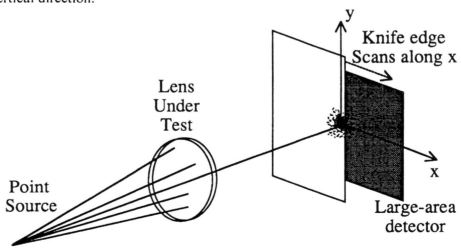

Figure 5.11 ESF test setup using a point source and a knife-edge receiver.

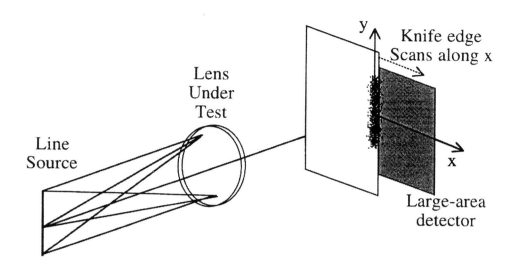

Figure 5.12 ESF test configuration using a slit source.

5.7.2 Averaging in pixelated detector arrays

The acquisition of PSF, LSF, and ESF data with a detector array has MTF implications in terms of pixel size and sampling as seen in Section 5.5. In addition, the use of a detector array facilitates data processing that can increase the SNR by averaging the signal over the row or column directions. Beginning with the PSF-test configuration seen in Fig. 5.13, we can sum PSF data along the y direction, which yields an LSF measurement in the x direction:

$$\text{LSF}(x_i) = \sum_{j=1}^{M} \text{PSF}(x_i, y_j) \ . \tag{5.18}$$

Summing the PSF data along the y direction and accumulating along the x direction yields an ESF measurement in the x direction of

$$\text{ESF}(x_{i'}) = \sum_{j=1}^{M} \sum_{i=1}^{i'} \text{PSF}(x_i, y_j) \ . \tag{5.19}$$

Because of signal averaging, the LSF and ESF test data of Eqs. (5.18) and (5.19) will have a better SNR than the original PSF test of Fig. 5.13.

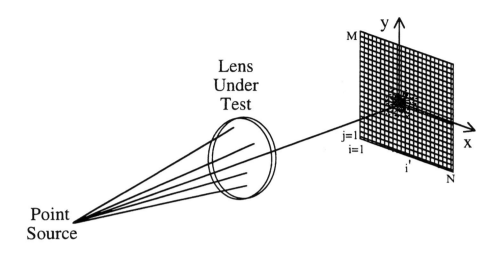

Figure 5.13 PSF test configuration using a two-dimensional detector array can be used to produce PSF, LSF, and ESF data.

Similarly, if we use a line source oriented along the y direction and sum the LSF data along y, we obtain an LSF measurement with better SNR. If we accumulate the LSF data along the x direction, we obtain an ESF measurement. If we then sum the ESF data along the y direction, we obtain an ESF measurement with better SNR.

5.7.3 Potential pitfalls

When using the signal-averaging techniques just described, we must be especially careful to subtract any background level in the data. Expressions such as Eqs. (5.18) and (5.19) assume that the detector data are just the image-plane flux. Also, if we sum over columns, we must ensure that each column has the same data; watch for an unintended in-plane angular tilt of a slit or edge source with respect to columns. Another item to consider is whether to window the data, that is, to use data only from the central image region where the flux level is highest and the projection optic has the best performance. If the lens under test has field-dependent aberrations, it is particularly important that we use data from the sensor-array region which contains the data from the FOV of interest.

5.8 MTF testing observations

The MTF measured by two different methods, for instance, a bar-target test and an impulse-response test, should be identical. When the measured MTF is not identical, some portion of the system (typically the detector array) is not responding in a linear fashion. Accuracy of commercial MTF measurement systems ranges from 5% to 10% in absolute MTF, with 1% accuracy possible. Remember that radiometric measurements are difficult to perform to the 1% level, and that MTF testing combines radiometric- and position-accuracy requirements. The position accuracy and repeatability requirements are often quite demanding – often less than 1 μm in many cases. Good micropositioners and solid mounts are required if we are to achieve high-accuracy MTF measurements. Even with the best components, lead-screw backlash can occur; high-accuracy motion must be made in one direction only.

5.9 Use of computers in MTF measurements

Most commercial systems come with computer systems for data acquisition, processing, and experimental control. Precise movement control of scanning apertures and other components is very convenient, especially for through-focus measurements, which are tedious if done manually. The graphics-display interface allows us to immediately visualize the data and lets us adjust parameters in real time. Digitized data sets facilitate further signal processing, storage, and comparison of measurement results.

There is, however, a drawback to the use of computers: software operations typically divide out instrumental MTF terms, using assumed values for the slit width, collimator MTF, wavelength range, $F/\#$, and other parameters. In practice, you must ensure that these settings are correct. It is good practice to check the final results for realism as discussed in Chapter 4.1. Also remember that a variety of spurious signals can trigger the electronics of the data-acquisition system. It is possible to take erroneous MTF measurements on a clock-feed-through waveform from the detector array electronics, rather than from the PSF of the system. As a test, cover the detector array, and see if an MTF readout is displayed on the monitor. Occasionally this does happen, and it is well worth checking if the measurement results are in question.

5.10 Representative instrument designs

We now consider four different instrument approaches that represent the state of the art in commercial MTF systems. We examine the design tradeoffs, and compare the systems for their viability in visible and infrared applications.

5.10.1 Example #1: visible ESF

In the visible ESF design (Fig. 5.14), intended for MTF measurement of visible-wavelength lenses, a pinhole is illuminated by either a laser or filtered arc lamp. A relatively high flux level and good SNR allows for a simple system. A tilt control for the lens under test allows for off-axis performance measurement. A diffraction-limited refractive collimator limits the aperture size of lenses that can be tested. Larger apertures could be obtained in reflective collimators. A scanning "fishtail" blocking device is used as a knife edge in the image plane. The data are thus ESF, even with a pinhole source. The configuration of the knife edge allows for a measurement in either the x or y direction without changing apparatus. A two-dimensional scan motion is required. A photomultiplier tube detector is used, which is consistent with the visible-wavelength response desired and which preserves the SNR of the instrument. The data-processing procedure involves a spatial derivative of the ESF and a numerical fast-Fourier transform (FFT) to produce the MTF. The derivative operation required to go from ESF to LSF is not problematic in terms of noise because the system has a high SNR.

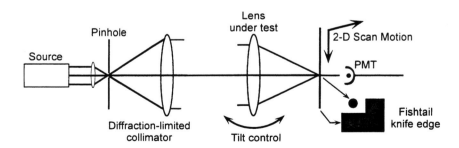

Figure 5.14 Example system #1: visible ESF.

5.10.2 Example #2: infrared line response

The infrared line response system (Fig. 5.15) was designed for LSF measurement of lenses in the infrared portion of the spectrum. A number of techniques were employed to enhance the SNR, which is always a concern at IR wavelengths. A heated ceramic glowbar was oriented perpendicular to the plane of the figure. It provided an IR line source, which gave more image-plane flux than a point source at the same temperature. A long linear scanning detector, oriented parallel to the source direction, was used to acquire the LSF image. The narrow detector width served the function of a scanning slit. The detector itself moved, so that there was no need for a movable cooled slit in addition to a cooled detector. A diffraction-limited reflective collimator was used, which allowed for larger UUT apertures, and broadband IR operation without chromatic aberrations. A chopper just following the source facilitated narrowband synchronous amplification, which reduced the noise bandwidth and increased the SNR. The LSF was measured directly. This approach avoided the derivative operation with its associated noise penalty that would have been required for an ESF scan, and also required less dynamic range from the sensor.

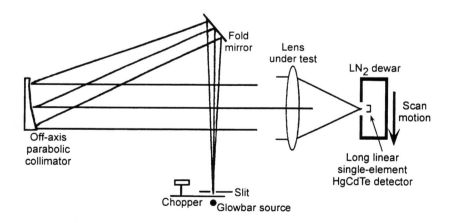

Figure 5.15 Example system #2: infrared line response.

5.10.3 Example #3: visible square-wave response

The visible square-response system[4] is rather special in that it is a true infinite square-wave test rather than a bar-target test. The system was configured for testing the MTF of lenses in the visible portion of the spectrum. The object generator seen in Fig. 5.16 served as the source of variable-frequency square waves. The object generator consisted of two parts, a radial grating which rotated at a constant angular velocity and a slit aperture. The center of rotation of the grating was set at an angle $(0 < \theta < 90°)$ with respect to the slit aperture. This allowed control of the spatial frequency of the (moving) square waves. The response of the system was then measured one frequency at a time.

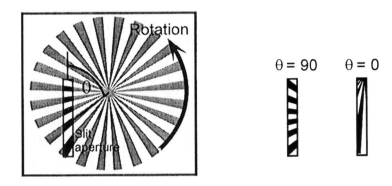

Figure 5.16 The object generator for the visible square-wave test.

The object generator was backlit and, as seen in Fig. 5.17, placed at the focus of a collimator and re-imaged by the lens under test. A long narrow detector was placed perpendicular to the orientation of the slit image, and the moving square waves passed across the detector in the direction of its narrow dimension. The analog voltage waveform out of the sensor was essentially the time-domain square-wave response of the system to whatever spatial frequency the object generator produced. To avoid the necessity of using the series conversion from CTF to MTF of Eq. (4.20), a tunable narrowband analog filter was used, which allowed only the fundamental frequency of the square wave to pass through the filter. The center frequency of the filter was changed when the object frequency changed, and all of the harmonics of the square wave were eliminated. The waveform after filtering was sinusoidal at the test frequency, and then we measured the modulation depth directly from this waveform. Filtering the fringes electronically required that the fundamental frequency was always well away from the $1/f$-noise region below 1 kHz, which implied a fast modulation in the object generator. This necessitated a relatively high power level from the optical source that backlit the object generator, because the detector had to operate with a short integration time to acquire the fast-moving fringes. This design worked well in the visible but was not as feasible in the infrared, where the SNR was lower and, consequently, a longer integration time would have been required.

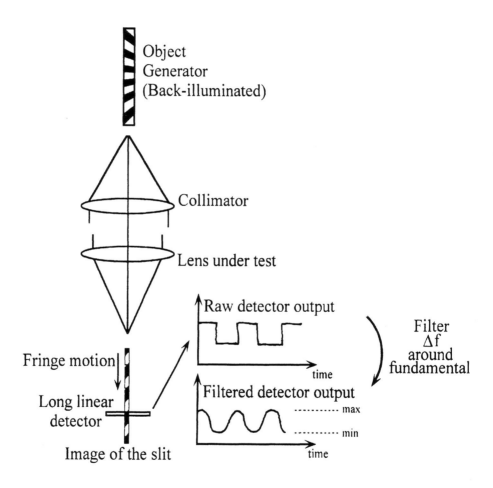

Figure 5.17 Example system #3: visible square-wave response.

5.10.4 Example #4: bar-target response

The IR bar-target system seen in Fig. 5.18 was intended to measure the bar-target response of lenses at infrared wavelengths. The four-bar patterns were made of metal and backlit with an extended blackbody source. A separate bar target was used for each fundamental spatial frequency of interest. The resulting image was scanned mechanically with a cooled single-element detector. Because the image stayed stationary in time, we were able to use a scanning speed slow enough to allow for signal integration, thus increasing the SNR. We could not use an analog narrowband filter, hence the data set required an analytic correction that converted the bar-target data to MTF. We could not electronically filter out the harmonics as in example system #3, because the slow scan speed would have put the resulting analog waveforms in the $1/f$-noise region at low frequencies.

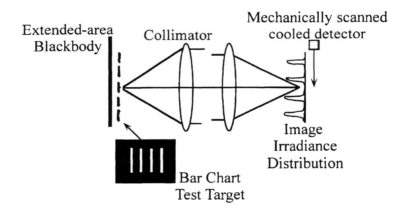

Figure 5.18 Example system #4: infrared bar-target response.

References

1. J. B. DeVelis and G. B. Parrent, "Transfer function for cascaded optical systems," *JOSA* 57, p. 1486 (1967).

2. T. Alexander, G. D. Boreman, A. Ducharme, and R. Rapp, "Point spread function and MTF characterization of the KHILS infrared-laser scene projector," *Proc. SPIE* 1969, pp. 270-284 (1993).

3. S. E. Reichenbach, S. K. Park, and R. Narayanswamy, "Characterizing digital image acquisition devices," *Opt. Eng.* 30, pp. 170-177 (1991).

4. L. Baker, "Automatic recording instrument for measuring optical transfer function," *Japanese J. Appl. Science* 4 (suppl. 1), pp. 146-152 (1965).

Further reading

L. R. Baker, *Selected Papers on Optical Transfer Function: Measurement*, SPIE Milestone Series, Vol. MS 59, 1992.

Index